THE Blind Tour Guide

THE Blind Tour Guide

Patrick Lambe

TIMES BOOKS INTERNATIONAL
Singapore • Kuala Lumpur

Published by Times International Publishing
under the imprint Times Books International
Times Centre, 1 New Industrial Road, Singapore 536196
Fax: (65) 62854871 Tel: (65) 62139288
Email: te@tpl.com.sg
Online bookstore: http://www.timesone.com.sg/te

First edition 2002

Printed in Singapore by Utopia Press Pte Ltd

National Library Board (Singapore) Cataloguing in Publication Data

Lambe, Patrick, 1960-
The Blind Tour Guide :Surviving and Prospering in the New Economy / Patrick Lambe. - Singapore : Times Books International, 2002.
p. cm.
ISBN : 981 232 367 8

1. Knowledge management. 2. Information society. 3. Technological innovations --Management. I. Title.

HC79.I55
303.4833 — dc21 SLS2002014803

Contents

Preface

This is a collection of chapters on global themes affecting knowledge workers around the world. It is about the challenges, risks and opportunities imposed by the phenomenon of being connected. And yet it is written from the ground in a country so small, so connected and so ambitious, that you can practically see the new economy take shape around you every day.

While old economy ostriches decry the internet boom as gone forever, and the economists discover that the numbers underpinning the original definition of the "new economy" were not what they thought they were, this book continues to use the term "new economy". In it, we attempt to explore the broader impact of technological change on business and working life. While academicians argue over definitions, the depth, breadth and pervasiveness of this particular set of changes cannot be denied.

Singapore is a unique city. Straddling Asian cultures and Western business ideas, it exhibits the best and the worst of both. What's more its economy is so highly tuned, that when the government decides that Singapore will become a knowledge based economy, then the entire country gears up and starts an incredible odyssey practically overnight.

It is easy to mock, but difficult to imitate such determination. It is not so easy to mock some of the products of that enterprise: its imminent transition to becoming the most connected society in the world, a workforce development programme that has borrowed generously from international experience, but which has in its sheer scale and integration surpassed any in its effectiveness, and an economy that emerged virtually unscathed out of the Asian meltdown of 1997-1999.

Singapore is, essentially, a laboratory where you can see the transitions between old and new economy happen every day, in human scale. You can practically watch people cross the street. And that is why I hope these essays, written from a Singaporean context and perspective, might in their very particularity find a wider audience and a broader relevance, representing one view of the way the world really works.

In bringing these essays together for publication, I have five key people to whom I should offer special thanks. This book is dedicated to Edgar and Mary, for looking after me, feeding me and allowing me to get a unique view of the world at ground level in Singapore. I would like to thank Anna Teo of *The Business Times* for encouraging me to write for her Knowledge Economy page – most of the chapters in this collection originally appeared on Anna's page. I would like to thank Liam Brown of the British Council for his unbounded generosity in comment, critique and a steady supply of always apposite reading matter. And finally, I would like to thank Sim Wong Hoo of Creative Technology, whose insights and achievements underpin the chapters on innovation.

Patrick Lambe
May 2002

1
Becoming a New Economy Manager

There is no single digital divide. And the digital divide is not simply a matter of the gap between technology haves and have-nots. Behind this simple description of the rift there are deeper and bleaker gaps in literacy, skill and attitude that few of us see - even in ourselves. While commonly described as a special problem with the lesser-educated older worker, these fault lines actually cross the entire spectrum of employment from operator to chief executive, and span the range of organisational forms. Failure to address these gaps will mean failure for workers in the market for employment and it will mean failure for organisations in the market for customers.

The digital divide is usually described as a socio-economic problem. At a global level, it suggests that less developed countries are increasingly being disadvantaged by their lack of access to technology and Information Technology (IT) mediated knowledge. In a global economy where a larger and larger share of business is knowledge business, and where even traditional economic activity is being changed through knowledge applications, the cards in the poor country's packs are steadily being depleted... the last one they can play being cheap labour. But supply chain management, increasing demands for "zero defect" quality processes, and IT systems integration along the whole value chain of production, means that even traditional low-skilled economic activity requires access to infrastructure and human skills beyond the reach of IT-poor countries.

To connect or not to connect

The debate rages on. Some point out that wireless communications technology enables poorer countries to leapfrog large, developed economies, because they have ageing infrastructures to maintain. Wireless internet technology is already here, and the communications backbone is the biggest single barrier to IT development in any economy.

Market forces can provide equipment, so long as the environment can support stable knowledge and information flows.

In Singapore during 1999, the lessons of recession, retrenchments and rapid economic change struck home hard. In the closing months of 1999, four ministerial speeches referred to the problem of the digital divide in Singapore. Singapore is currently the fourth most connected economy in the world, and is likely to be number two within two years, and every Singaporean student is being taught IT and internet skills in school. But Mr Lim Swee Say, Minister-of-State for Communications and Information Technology also pointed out that the number of non-connected Singaporeans outweighs the connected by three to one. A Channel NewsAsia Gallup poll conducted in March 2000 found that almost half of all Singaporeans were worried about their ability to keep up with technology. Singaporean efforts to overcome these problems are massive, precise, and coordinated, from physical infrastructure to "dot comming" all government services, to training and learning initiatives at every level of the population.

The key issue of the digital divide in the Singapore context is seen as the gap between the younger and higher educated workers, versus the older and lower educated workers. These are the key numbers that strike home:

- Three quarters of Singaporeans are not connected to the internet
- Two thirds of the Singapore workforce did not progress beyond secondary education
- Three quarters of the Singapore workforce are above the age of thirty - the vast bulk of their education and work experience and attitudes are pre-internet

What does this mean for business and the workforce? Simple access to IT is only the tip of the problem. The generational divide goes far beyond educational level attained. Here is a key, disturbing fact: almost all Singaporean senior managers, vice presidents, chief executives, grassroots leaders and senior civil servants are pre-internet in formation and experience. Singapore is not, of course, alone in this - it's a global phenomenon.

A global problem

Digital illiteracy is rampant among leaders in every walk of life, in every developed economy in the world. Managers are having to work hard to keep up with their juniors. In General Electric's European operations, senior managers have introduced "reverse mentoring" where senior managers can increase their knowledge of the internet and e-commerce by meeting every week with a younger person on their staff who is an expert on the subject.

> British Prime Minister Tony Blair told a self-deprecating anecdote about himself at the Davos Economic Forum in 2000. He had decided to overcome his own computer phobia and enrolled for an IT course in his constituency's community centre. At the final examination, he asked a nervous young man next to him if he was so tense because he was next to the PM. "No," replied the youth, "I'm tense because I can do this stuff and I'm unemployed, you can't do it, and you are Prime Minister."

Behind the humour of this story lies a disturbing perception. Working life at every level is seeing the most radical shift since the transition from agricultural economies to factory economies in the industrial revolution – but the impact will be more profound because it is not a physical transition from one type of workplace to another, but a rapid, discontinuous series of shifts within the same physical and organisational space.

The tension between the rising economic power of the digitally literate junior worker, and the increasing irrelevance of the digitally illiterate manager, can only be played out painlessly through entrepreneurial departures from the traditional firm and youth-weighted dot.com ventures. With the dot.com crash, even that avenue has disappeared. Like an enormous continental shift that must work its way to the surface in seismic shocks, workplaces, companies and entire industries are experiencing the impacts of these discontinuities through manager and employee frustration, poorly understood and poorly implemented change, resistance to change, internal conflict, poor communications. Dot.com companies translate these rifts to the marketplace, where the new rules of battle sweep away the old paradigms of doing business, whether successful or not.

The Net generation worker	The pre-Net generation manager
• demands immediacy	• mistrusts haste
• ignores boundaries	• relies on boundaries
• values independence	• relies on dependencies
• values openness and curiosity	• is trained to consider risk, confidentiality, and "need to know"
• interprets 100 emails a day as "being in the loop"	• interprets 100 emails a day as information overload
• is young, lacks life experience	• is experienced, relates to the big picture
• is intolerant of adversity	• has learned how to work with and sometimes leverage on adversity
• throws energy into projects	• rations energy
• is prone to attention deficit, multitasks easily, lacks sustained analytical and reflection skills	• prefers mono-tasking, is dissatisfied with a lack of focus and time to devote sustained attention to issues
• collaborates indiscriminately	• collaborates selectively
• collaborates with peers	• competes with peers
• values respect from peers above all others	• values respect from superiors above all others
• knows how to authenticate people's trustworthiness, gives trust	• is conditioned against personalising work relationships
• gives respect based on performance (by own standards)	• gives respect based on status and performance (based on corporate standards)
• innovates	• standardises
• values questions	• values answers
• sees self-learning as central	• sees self-learning as an add-on
• mistrusts and challenges systems	• values and defends systems
• is self-directed	• looks to superiors for direction

But it is not a simple matter of throwing away old ideas and picking up new ones. Net generation workers are very different creatures in perception, attitude, skill and competency from pre-Net generation workers and leaders.

Some aspects of this can be ascribed to every generational gap between the young Turks and the old hands since time began, and like all generalisations, there are counter-examples in both camps. You can have self-directed creative managers and you can have risk averse Net-generation workers. However, even the most cursory comparison of internet relay chat discussions with corporate small-talk points up key differences in generational attitudes towards innovation, relationships, communication and personal values. Understanding the differences throws a new light on the workplace experiences of many managers and employees.

2
How are the Mighty Fallen

> In ancient Nordic legend, when warrior heroes die they go to Valhalla, a place of great feasting, golden mead and limitless food. After sating themselves on wine, women and song, they turn to their favourite sport, mortal combat. They fight till only one is left standing. Yet in the morning they rise again, whole and complete, to start their heavenly round of feeding and fighting again.

In March 2000, Chief Executives were resigning, retiring or getting fired at a pace of three per business day. By June, according to outplacement consultants Challenger, Gray & Christmas, the haemorrhaging had doubled, to an average of six per business day. In the first six months of 2000 alone, we saw Dirk Jager (Procter & Gamble), Rick Thoman (Xerox), John McCoy (Bank One), John Reed (Citigroup), Dale Morrison (Campbell Soup), Bob Ayling (British Airways), Jill Barad (Mattel), George Fisher (Kodak Eastman), Hasso Plattner (Baan), Dino Adriano (Sainsbury), and Douglas Ivester (Coca Cola) bite the dust.

From diamond mining to canned soup, from insurance to software, from toys to supermarkets, from banking to waste management, the corporate world is seeing a bloodletting unprecedented in its breadth and dimensions. Like the Viking warriors, these heroes will rise to fight again as CEOs of different companies, considerably richer than before. Yet the same blithe shrugging off of wounds and damage to reputation cannot be ascribed to the battered organisations they have left behind.

> Barclays Bank and Reed Elsevier took over a year to find suitable successors to the CEOs they removed, Marks and Spencer drifted for over six months, and the troubled software company Baan has had more CEOs than press releases can easily reconstruct in the four years since it ousted founder Jan Baan from the post.

In 1997, management thinkers were writing cogently of the need for careful succession planning in large corporations, and indeed several good models emerged from the likes of General Electric and Coca Cola. By 1998, there was concern that the average tenure of CEOs had declined to four years, and was falling. Of my year 2000 hitlist, half had been in their jobs for less than two years. Corporate leadership consultants Warren Bennis and James O'Toole have now coined the term "CEO churning" to describe a pattern of turmoil and disorientation in the companies they leave behind (*Harvard Business Review* May-June 2000)

The interim CEO

CEOs are sacked for lots of reasons: the pace of change is too slow; the pace of change is too fast; resistance from the middle managers and corporate culture; differences on strategy with increasingly assertive shareholders and board members; not being tough enough; being too tough; botching expensive acquisitions and mergers; financial mismanagement.

Of the high profile firings over the two past years, common themes emerge however. CEOs of established companies are finding it harder to move their organisations into prosperity growth as fast as the markets they are competing in. Companies are being acquired without following through on structural and cultural change. Mergers are completed without following through on communications and culture issues. Old economy CEOs used to autonomy and tough leadership styles are increasingly confronted by angry managers and shareholders who require greater accountability, more transparent information and communications. Everybody wants the CEO, suddenly, to listen to them. Not to mention the gut wrenching speed of technological change and transformed business models wrought by technology and mushrooming startups. Botched management of change, and botched communications underpin the failures to percolate necessary change through the organisation.

The risks of being a CEO are higher, of course. Nobody knows how long they will last, and boardroom coups engineer a nervous atmosphere of intrigue, mistrust and cloaked communications reminiscent of the worst aspects of Imperial Rome.

New CEOs require costly "prenuptial agreements" giving them ever higher payoffs should they be given the shove before their time is properly up. Jill Barad of Mattel walked away with US$40 million.

And good CEOs are ever harder to find; not simply because they are licking their wounds from the last lost battle, but because the skills sets needed now are ever more complex. Troubled companies are increasingly turning to "interim CEOs" to play the necessary bad cop and take on nasty corporate makeovers, before appointing more permanent "good cop" bosses.

Ousted CEOs live to rule again, it appears, after a few months' break. The companies they leave behind suffer more, because the issues they grapple with daily never go away. The new economy tensions have started to jettison leaders far more easily than is healthy, and these tensions continue to ripple through organisations in waves of conflict, stress and disorientation.

The new economy leader

Most of us, apart from the most lucky, work in a divided, dysfunctional workplace. Our corporate and organisational structures are built on the principles of command, coordination and control, and so are our internal investments in systems and processes. As leaders and managers we play lip service to the rhetorics of empowerment, lateral communication and team synergies, but we maintain our reporting, decision-making, management information and authority systems in an uneasy parallel continuum, seeking to get the best out of both worlds, refusing to accept that they may be mutually exclusive.

The schizophrenia of talking up the empowerment of our human capital while systematically reducing our dependence on increasingly disloyal individual knowledge workers through IT/IS knowledge capture systems, for example, is rarely recognised. We ask our employees to invest extra commitment in us and promise in exchange an increased marketability in a rapidly evolving employment marketplace, but we fail to acknowledge that our promises are always in an indefinite future, and we too rarely provide for increased marketability within our own organisations.

Our management investments are almost always in systems for coordination and control, whether human or technical, and most CEOs we recruit reinforce these principles to deadly effect. We invest negligible amounts in the influencing of human culture within our workplaces, and in the influencing of human attitudes and behaviours, despite the swathes of evidence that demonstrate the costs of ignoring them, and the enormous benefits gained by organisations that have invested in them.

We lack the courage to invest in the human, and we comfort ourselves with a rhetoric that persuades us we have done so. To take just one example of this, review the executive appointments section of the newspaper and look for Knowledge Manager, Network Manager, IS Manager, Human Resource Manager posts. They are all posts which have well-defined technical and professional requirements, as well as a proven track record of experience and achievement. Nine times out of ten, you will find within the key requirements of these posts "good interpersonal, teamworking, and communication skills", up there with C++ and XML programming, TCP/IP, object oriented methodologies, manpower planning, compensation/benefits and industrial relations. Increasingly, the ability to manage change is being incorporated into such requirements. And it is right and proper that such qualities should be recognised.

The schizophrenia lies in this: while there are public benchmarks for measuring C++ and payroll management abilities, we have no standardised way at all of assessing a person's interpersonal communication or change management skills, beyond subjective testimony through referees and our own impressions at interview. Some organisations use psychometric testing to persuade themselves that they are measuring something, and despite the dubious links between a person's interpersonal abilities and their performance in a test, at least these organisations are attempting to measure something. Most of us do not even attempt to achieve such a measure.

Perpetuating doublethink

The Singapore PSB's national certification programme called Critical Enabling Skills Training (CREST) is the first time a broad range of soft skills critical to the knowledge-based workplace has been established as a standard of achievement and performance. The programme is now almost four years old, and not one of the employers that asks for "good interpersonal communication and teamworking skills" has asked for evidence through CREST certification.

So there it is: we recognise these soft skills as core to the requirements of a post, heck, we pay for the extra space to put them into our advertisements, up there with the four other most important criteria for appointment. But we neither measure objectively nor invest substantially in making these qualities central to our workplace. We perpetuate doublethink in our organisation, so that even while sending our frontline workers on communication skills training courses, we pursue filtering and "need to know" policies at higher levels. Good training managers know that the most common refrain from a successful interpersonal skills course for front-line workers is "Why don't you send my manager on this course?"

We know these skills are important. We know they are good for us and our organisation. But the very fact that we don't know how to measure the impact or effect of such skills training demonstrates that there is a radical discontinuity between how our organisation actually works, and how we prepare ourselves to work for it. Worse, we do not know how to define the qualities we really need from our leaders. Gone is the day of the macho, mead swilling corporate warrior CEO. Let's have some sense here, and a more humble generation of leaders who recognise their limitations yet know how to exploit the strengths of others.

3
Sick Companies

In my first chapter I argued that there was a deeper digital divide between old economy and new economy workers than simply the ability to manipulate technology. Subsequently (in Chapter 2), I laid part of the blame for our discontinuities on the failure of leaders to adopt new models of leadership more appropriate to the transitions we must undergo. In this chapter I want to focus on that other, more difficult area of resistance to change: the interplay between leadership and corporate culture.

> When European colonists populated the Pacific Islands, smallpox, syphilis and the common cold laid the native populations waste. The diseases were unknown to them, and they had no natural defenses.

In the same way, established old economy businesses have defensive reactions to these new economy viruses at very visceral levels, far below the rational and strategic level most managers profess to employ. We can identify just three of the many syndromes that companies suffer from, to one degree or another: *The Component Syndrome; The Ceaucescu Syndrome; The Forbidden City Syndrome*.

The Component Syndrome

The *Component Syndrome* arises from a mechanistic rather than an organic or ecological view of organisational working. This view states that jobs and roles are mapped, structured and periodically reconfigured within the best system design for the business being engaged in. Specifications for the individual jobs within the structures are drawn up in a very similar way to machine design, and personnel recruitment is treated very much like a component sourcing exercise. Either appropriately configured humans are sourced on the job market through a system of testing against specifications, or they are engineered into the jobs through internal employee development programmes.

Training and development is centred around employee sculpting, either to create best fit for a predefined job, or to do corrective engineering when there is poor fit against specifications. Increasingly organisational theorists are exploring organic or ecological models for how organisations work, because the mechanistic model is largely unattractive to individual human aspirations and ideals of personal growth and development. A recent article in the *Harvard Business Review*, for example, explored the novel notion of "job sculpting", where an individual is encouraged to mould the job according to his or her personal long term goals and aspirations as well as organisational goals (*Harvard Business Review* Sept-Oct 1999).

There are, however, few organisations untouched by the mechanistic model, and there are good reasons for this. Systems that treat employees like replaceable components appear to reduce the risk of loss through employee departure. Performance that is individuated to individual skills and abilities increases difficulties in management, consistency and replacement. As knowledge workers become more mobile, informed and market-savvy, loyalty and long-term allegiance declines. They are not only more aware of the opportunities beyond their present post; they also remember the vast corporate betrayal of the downsizing exercises of the 1990s.

It is hard to resist depersonalising and de-individuating your systems and structures, because it appears to reduce your level of risk in running a consistent and efficient workflow system. And yet, at the same time, we are in an environment where component-based production of tangibles is rapidly losing ground to intangible knowledge-based saturation of work processes at every level in the organisation. Creativity, ideas, questions, learning, improvisation, innovation, decision-making, all need to happen much faster and closer to the front-line in internet-time, if the organisation is going to prosper or even survive.

To create or encourage this knowledge-activity environment, is to unleash the power of the individual over your systems, and this is where the radical discontinuity of the Component Syndrome lies. The dilemma facing the corporate mind appears almost impossible, because it appears that you cannot survive unless you standardise and depersonalise your systems and knowledge flows, and yet it also appears that you cannot survive unless you encourage in your employees skills and attitudes that challenge those very systems.

The Ceaucescu Syndrome

Nicolae Ceaucescu was President of Romania from 1965 to 1989. His regime was characterised by communist-inspired rigidities of structure and control, and probably the world's most pervasive citizen surveillance system, with (the story goes) every telephone manufactured being issued with a bugging device as standard. Enforced resident relocations into concrete housing blocks to make way for grandiose monuments and megalomaniacal building projects, rigid control of industrial output and increasingly xenophobic alienation from the external environment created immense tensions within Romanian society, and the greater the tensions, the higher the level of repression and standardisation and control. In late 1989, the growing popular unrest was suddenly unleashed when the army abandoned Ceaucescu, and within six days of a popular uprising in a regional city, the most powerful dictator in Europe was overthrown in a bloody coup that claimed his life and that of his hated wife Elena.

Employee surveillance, whether through hidden cameras, software that monitors email and internet usage, or activity based costing, demonstrate, to differing degrees, features of Ceaucescu Syndrome, as do apparently pathological attachments to restructuring and re-engineering exercises, and single-minded pursuit of prestige corporate projects at the expense of employee interests and development. Organisations which suffer from this syndrome typically combine highly developed management information systems geared around imposed targets, with low levels of trust and communication. Productivity is defined according to internally driven standards, and attitudes towards competitors are aggressive and highly negative.

It is easy to understand why Nicolae Ceaucescu became progressively more megalomaniacal the more he became divorced from the realities of his economic and political environment. The only way he could combat the threats to his power either externally or internally was to step up the levels of control, and to impose his own view of reality by force. His prestige projects were frightening attempts to shape the external world according to his own, highly alienated vision of the world. In the end, the real world won, with bloody consequences for the dictator.

Weak forms of Ceaucescu Syndrome are highly pervasive. They

arise wherever you have managers whose perceptions, competencies and skills are inadequate to new challenges and demands, either from their subordinates, peers, or the wider environment. Typical reactions involve greater emphasis upon monitoring and control, over-defined work processes, defensive and paranoid reactions to anything that might demonstrate their weakness, and an intolerance for variations from the published norm. Alongside this, but almost completely disengaged from it, there is a highly developed rhetoric of leadership and vision, idealising model behaviour and demonising deviant behaviour.

The syndrome is, of course, highly dangerous for organisations and for individuals, as it was to Ceaucescu himself. The defensive response to a challenging environment is natural when there is a relationship of power at stake. Unfortunately, the easiest way to do this, if you are in power, is to redefine reality according to the old power relationships. Leaders, of course, whether they be political or organisational, have always controlled and defined communication and information flows. We are all practised in requesting and propagating pre-defined information that supports our position, and in demeaning the importance of information that undermines our position. And if it is true that organisations become what they measure, we do indeed get to define who and what we are.

Unfortunately, the world is a good deal bigger than our perceptions of it, and unless we, as leaders, can accept the challenges and sometimes painful reality checks of our environment, the discontinuities proliferate, suppressed only by means of increasingly aggressive propaganda, until the control system fractures and collapses under the pressures of the real world. Ironically, it is exactly those systems and instruments of information supervision and control that lead to our downfall. Networked corporate email systems were installed on the back of a business case for faster and more timely management information flow vertically within the organisation. Their impact has been chaotic, boundaryless and highly democratic lateral knowledge flow within organisations. Distorted or propagandistic perceptions of reality are now harder and harder to impose, simply because people can communicate with each other more easily.

There is another irony here. We treat the rhetoric of empowerment, creativity and innovation as just another rhetoric, a tool to persuade individuals to align themselves to our goals. In order to protect ourselves from irrelevance and uncertainty, we do not align our own working practices and organisational systems with this rhetoric, because we do not know how successfully we ourselves can compete in an empowered, creative and innovative workplace. So for the sake of rhetoric we invest in training our workforce in all the right knowledge economy skills, but for the sake of our security we maintain our old systems of control, decision-making and measurement. The danger to us is that the generations behind us do not see these skills as a rhetoric, but as a reality, and they are skills that will indeed empower them to replace us.

The Forbidden City Syndrome

The end of the Qing dynasty, with a ruling elite paralysed, frozen in time, and apparently immune from the vast tidal forces of change sweeping the Chinese Empire, is powerfully symbolised by the walls and the rituals of the Forbidden City.

The ruler who is removed from direct access to knowledge of his empire is one of the most persistent images of human history. The stories in the Arabian Nights of the king who goes among his subjects in disguise to find out the truth about his kingdom finds modern parallels in the exploits of King Abdullah of Jordan, and in press speculation about the legitimacy and dangers of the spin doctors and special advisors in Prime Minister Blair's government, or former President Wahid's perceptions of what was happening in Indonesia.

And yet, the more senior we are in organisations, the more systematically we filter and define the information we want to see about the external world.

> A five year research project by the Jenson Group in the United States recently found that 80% of front-line knowledge workers in the United States feel that they do not have sufficient information navigation and handling skills to make effective decisions in their jobs. Only 60% of their managers felt the same way.

The research did not survey senior executives, but we can suppose that the levels of confidence grow the higher you go. There are two possible explanations for this, and they are not mutually exclusive. The first could simply be that the more effective you are in finding and processing information, the more likely you are to be promoted. The second, equally valid explanation says that the higher you are, the more you get to define what you see. While the front line must deal with a cacophany of information "noise" and few resources to filter it, most of our management information systems are designed to filter out the noise and produce measures of reality that our rulers can use to make decisions and formulate their strategies.

This works fine in a relatively stable environment, punctuated periodically by significant change. In a networked environment where change is a series of chain reactions going off all around like strings of firecrackers, where new business models reverse strategy and originate it from the consumer end of the value chain, the Forbidden City Syndrome is a severe liability. The knowledge that counts is the knowledge buried in the noise at the front line of our organisations. The walls and series of gates, the filters and stages of purification of our knowledge as it progresses up the management information chain to the Chief Executive, increasingly isolate the strategy of the mature organisation from the real world activity outside.

Chief Executives who do not have access to unstructured information and chaotic noise, will never develop the skills to navigate it and use it. Organisations which do not transmit information skills to their front line will never absorb market-critical knowledge in time to adapt to the new environment. But transmitting the skills to the front line is not enough. The prospectors for gold at the front-line are no good at all, if the information flows towards informing strategy don't work. In the knowledge age, knowledge and information filters need to be defined at the front line, if the strategy makers are ever to stay in touch with the speed of change in the real world.

When we define what we see, we see only what we define. It is hard, but necessary, to learn how to see with another person's eyes, and we who are distant and protected from the front line need to learn that ability as a matter of urgency.

The discontinuity this introduces into working is related directly to our inability to understand what is happening at the edges of our organisations. At the outer skin of our organisations the amount of exposure to different options, amount of information, communication interactions, and decision-making needs, is currently increasing by factors of between two and five every three years. We don't see that. What we see is increasing dysfunction at our front lines – if we are enlightened, we interpret it as a skills gap, and throw training at it. If we are not enlightened, we fail to see the skills issue and we interpret it as an attitude problem. The demands of knowledge work, and our poor skills provision for it, place increasing stresses and pressures at the edges of our organisations, and radiate inwards. We are the last to feel them.

The organisations that survive will be those that resource the front line, and allow the front line to help define how the organisation sees the outside world. Training is not enough – systems must also follow, if what is perceived at the front line is to be communicated effectively into strategy.

4
Leadership in the New Economy

Imagine, if you will, that you are arriving for the first time in Singapore on a special holiday package, with your own individual tour guide as part of the package. You arrive at the airport, and there, as promised, is your tour guide. He is tall, about 50 years old, and very distinguished looking, but he is almost completely blind. He first started going blind in 1968, the year he graduated from tour guiding school, the year of flower power in California and student revolts in Paris, before man ever got to the Moon. Before you gird up your loins for an assault on your tour operator for short-changing you, imagine two more small things. First, you live in a world where all tour guides start going blind as soon as they graduate, and second, you live in a world that accords great respect and deference to tour guides. You never ever question a tour guide's ability to show you around.

In what circumstances could this tour guide be helpful to you in getting to know the fine city of Singapore? He has not seen most of its current landmarks, and his geography of the island will only be hazy at best. Our traditional image of the tour guide's role as being one of lead, point and talk will be severely challenged. Perhaps, we venture, this tour guide could be useful at least in telling us about how Singapore used to be, and we could lead our guide, look and listen, and imagine a Singapore long gone by. We still feel short changed, but heck, tour guides deserve all the respect they can get.

Reflecting further, we find that we must talk to this tour guide. We must tell him what we see, in order for him to illuminate us as to the history of the place we are in. We suddenly find him adept at listening, and at questioning us. We are describing a visually rich environment to a blind man, and we find that we are acquiring a sharper eye for detail, a deeper perception of what we see as we describe it for him. And in his questions and his responses, we find that he is sharing with us the perceptions and reactions of 34 years worth of visitors to this spot. His blindness, quite suddenly, seems to be an asset.

Visual media revolution

The tour guide who starts going blind as soon as he graduates, is, of course, the manager and corporate leader. In an age of discontinuity and change, the metaphor of blindness does not seem too extreme. We live in an age where kids know more and learn faster from knowledge-based infrastructure, than their teachers do. Many adults will remember their first struggles with learning new hand-eye motor coordination skills in manipulating a mouse – many more adults, senior executives among them, have not yet mastered this basic skill. Children suckled on Nintendo and computer games look at them in wonderment as they struggle and curse like clumsy bears in a lightweight environment that seems totally natural to them. The visual literacy of students weaned on MTV surpasses by far our own.

I am 42. I still remember being disconcerted at the age of eight by our family's first television set – the year, strangely enough, was 1968. The woman getting out of the car disappears and in a flash is sitting at a restaurant table eating dinner, and I am confused because my whole life long so far, the world has revolved around me in long, continuous panning and zooming sequences. The flickering between man's face and woman's face at dinner, fragmentary shots of restaurant scenes that do not satisfy my curiosity, and inexplicable leaps in space time, leave me bewildered and faintly seasick.

Even the MTV generation is becoming old hat as the visual media become faster and more fragmentary. Kids of today can "read" information and messages off these new media better and faster than even 20-year olds, and my hat has long been thrown into the ring. I cannot compete.

In 1971, it was considered revolutionary to cut TV commercials to a mere 30 seconds. Today in Singapore, we average anything between 10 and 20 seconds per commercial, and that is considered pedestrian. Those who are in the buzz think that four to eight second commercials are the most hip.

Computer-literate youngsters read differently from book-literate managers. Screen reading is butterfly-like, flitting and scanning the information surface, picking out highlights, and dwelling momentarily on details. Book reading is linear, continuous, and invites rational chaining and sequencing of thoughts, more sustained attention and engaging deeper thinking processes. Small wonder that kids appear better at multitasking, faster at juggling thoughts in new and creative ways, but appear pathologically prone to attention deficit disorder, and weaker in consistency or original thinking skills.

New media introduced to children in their most formative pre-school years, produce generations that are the exact opposite of ourselves: trading our old economy capacity for deep concentration in exchange for a new economy mental agility and speed, and trading a preference for structured and pragmatic solutions in exchange for unstructured and creative approaches to problems. The landscape they inhabit is largely invisible to us.

We should not be surprised at this. Every generation in history has entrusted the incubators of the new world order to the guardians of the old world order. We have always inhibited our children by not seeing their world as well as they do. The speed of change in the last twenty years has been such, however, that the discontinuities and absurdities strike us with ever greater force. Never before has it seemed quite so obvious, that the student's capacity to learn far exceeds the teacher's capacity to teach, and the manager's capacity for leadership is sometimes sorely challenged by the drive and vision of the managed.

A paradigm shift?

We don't often acknowledge the depth of this rift. We talk of "changing mindsets" as if it were a matter of pressing a button. The general idea is that we need a paradigm shift to move our mindset up into a new gear, and all we have to do is think a bit harder and peer a bit harder at our surroundings, and the paradigm shift will happen naturally. As with many clichés, this is a distortion of the original concept.

> The term "paradigm shift" was most famously developed by the historian and philosopher of science, Thomas Kuhn, in his landmark book, *The Structure of Scientific Revolutions*. Kuhn did not describe the paradigm shift primarily as an *intellectual* shift, but rather, a *social* shift. Scientists build theories, or paradigms, he argued, based on features of the landscape of knowledge that they can

> identify around them. Once these key features become the substructure for the theory, they become dogmas. Sooner or later, new generations begin to notice features of the landscape that don't fit the theory. As more of them become noticed, the discontinuities with the ruling paradigm prompt new observers to propose a new theory.

Interestingly, Kuhn notes, very often the key pillars of the old theory completely disappear in the new paradigm. The vocabulary, the important features of the landscape are completely different. Both parties are looking at the landscape and see completely different things – what is visible to one party is invisible to the other – and because the main pillars of a paradigm form its substructure, people become dogmatic and aggressive, and violent discord often results. Differences in perception and vocabulary means breakdown in communication, and it means conflict. New paradigms gain force most often, therefore, not through intellectual discussion and dialogue, because communication cannot easily bridge completely different ways of looking at the world. New paradigms gain force through a process of natural selection: they work better than the old ones, more people grow into them and prosper because they work better, adherents to the old paradigms do not prosper, they multiply slower, they diminish and wither away.

Translated to the shift from task-based work to knowledge-based work, this is a bleak picture. It does not accommodate the vision and leadership of many older leaders, managers and politicians in driving the digital economy forward. It does not accommodate the levels of digital literacy that many pre-net generation workers and managers have achieved. Undoubtedly, it is possible to convert at least partially to new paradigms and new ways of doing things. Singapore's success and continuing economic power alone is witness to that.

In life, we know, no distinctions are absolute. However, in looking at the big sweeps of history, you have to look at the big numbers, and you have to acknowledge the pressures and anxieties that really exist. Many people feel the pressure of events beyond their control. Others feel excited by landscapes of opportunity. Both groups are looking at the same landscape. Those of us who were educated and formed in the pre-internet world will never see the full potential of the landscapes that polytechnic and university students now see. We can learn, sometimes painfully, to see the main outlines, but the future that is developing now does not belong to us, and no amount of unaided intellectual effort will deliver it to us.

Managing in the new paradigm

Let us return to our encounter with the blind tour guide. So long as we held him to the stereotype role of lead, point and talk, he was of little use to us. It was only when he listened to us, questioned us about what we saw, and shared his experience of dialogue with a generation of visitors, that his blindness became an asset that sharpened our awareness of what was around us. So it is with managers. Systems, rule books and personnel management are the remnants of an age when managers truly did know more than their employees. We now need managers who can draw out the ability of our colleagues to articulate what they see and experience, question them and help them shape it into meaningful strategy, compel them to develop the skills of deeper thought and sustained attention that they lack, and validate their sense of discovery and ownership of what they find. In this world, ignorance is a distinct advantage because only through ignorance can we frame the questions that will chart the steps of the new landscape

We are indeed at the cusp of a paradigm shift in perceptions of how economic and social reality works. What the Emperor Syndrome demonstrates is the vulnerability of traditional, old economy organisations, where perceptions are determined by old economy sensibilities, and information is shaped and moulded according to old economy filters. It is no wonder our workplaces are fractured, chaotic, confusing and frustrating. The Ceaucescu Syndrome suggests the dangers of resistance to externally imposed change, and the double-edged sword of a rhetoric of empowerment for our workers. And the Component Syndrome illustrates the extent to which old economy models still hold sway, and the extent to which new economy skills need to pervade senior levels of management as much as the front lines.

In these chapters above all, I have suggested that old economy trained leaders need to adopt a more humble, searching, questioning stance towards the newer generations, and accept that they have both strengths and weaknesses in the invisible landscapes of the new economy. The quality of our response to this new world depends on the quality of our relationships with those who see it most clearly. The wisest of pre-Net leaders are those who accept their blindness and seek out the most visionary of the new generation.

> Singapore's Senior Minister Lee Kuan Yew put it most aptly in his March 2000 address to the Singapore Techventure 2000 dinner in Silicon Valley: "I am not saying my mindset isn't from another generation, but I have young associates to remind me I am in the here and now."

5
The Panopticon

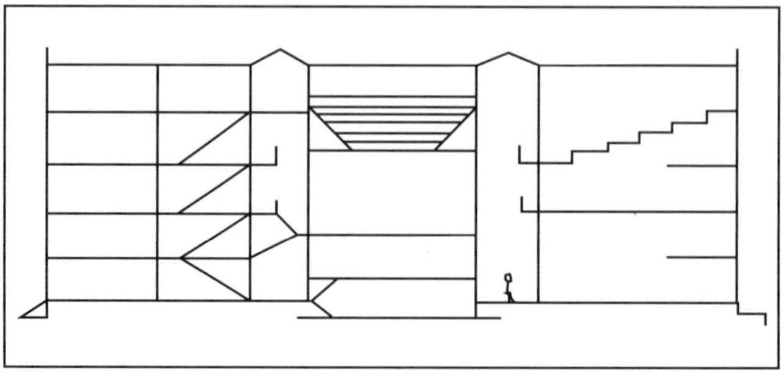

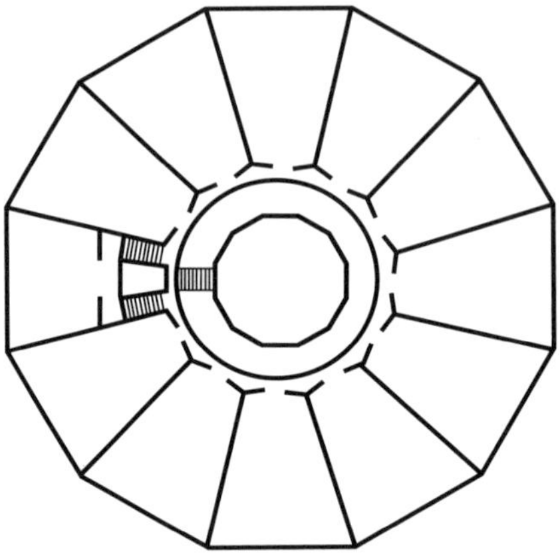

In 1791 the 18th century social theorist Jeremy Bentham published a new design for a prison. He called it the Panopticon. The design was such that the cells were arranged in a circle around a central space. The cell openings were arranged such that the prison warders could see the interiors of every cell at a single sweep, simply by turning their heads. The twist was that no inmate could see any other inmates; they could only see the warder at the centre.

It took almost 200 years for someone (the French social theorist Michel Foucault) to notice that hospitals, schools and factories have been organised on the same basic principles as prisons, and the reason why hospitals, schools and factories all feel alike, is because they all feel like prisons of sorts.

Since the groundbreaking work of Foucault, it has often been remarked that the organisation of prisons, schools, factories and companies, is built around the monastic division of labour and activity, powerfully symbolised by the individual human being isolated into a cell.

Radical human resource theorists such as Barbara Townley have launched blistering critiques on the walling-in and enclosure of talent within corporations, and the negative effects of such practices in terms of human development as well as corporate development. (*Reframing Human Resource Management*, 1994)

Whatever the politics of the case, the division of labour and the isolation of the individual in a highly defined cell, symbolised by the job description or the educational syllabus and final examination, are powerfully hierarchical models. As in Bentham's Panopticon, communication is strictly defined, and goes only upwards or downwards in the organisation in clearly defined formats. Work (or educational content) is standardised and distributed from the centre. Nobody talks to anybody else without permission - at least if they do, they are punished, one way or another. There are no rewards for being different. (Michel Foucault, *Power/Knowledge,* 1980, p. 228).

Walls cannot contain ideas

The problem with the networked society of the new economy is that networks make walls invisible. Where our old economy structures are all about command, containment, classification, standardisation, silence and systems, the new economy structures are about diffusion, broadcasting, noise, convergence, blurring of boundaries, confusion, high mobility of money, capital and people, and *networks*. The good news is that the effects are seen most where it comes to the movement of information, knowledge and ideas. You can still constrain physical assets by setting an armed guard on it, or building a physical wall.

Hence, this new scenario is not such a problem for prisons, where you do physically want to contain real human beings. It is a limited problem for manufacturing companies, which still do need to divide up the tasks of labour, and impose strong management controls. The new economy is much more of a problem for a growing number of companies, especially the ones that depend on selling knowledge and expertise, or the ones (like car companies) that have to make their products smarter by designing knowledge into them. It is a major problem for educationists, schools and universities, because their business *is* the communication of ideas, information and knowledge.

Now if you think about education just as a supply pipe into the workforce, radical changes in the needs of the work economy in terms of skills requirements, form a powerful suction effect on the school system. Our economy has changed almost overnight from a manufacturing economy into an information and services economy. We cannot keep churning out the same students, trained for highly defined jobs where all you had to know was in the manual.

Knowledge companies simply do not – *cannot* – work on the Panopticon model. Walls cannot contain ideas. New economy workers cannot behave like they are in a prison asking for permission to do anything non-standard. Once the walls are gone, the only model that

Information Skills	**Communication Skills**	**Networking Skills**
Navigating	Listening to Nuances	Mobility & Flexibility
Hunting & Gathering	Questioning	Adaptability to new environments
Filtering	Empathetic cyber-communicating	Openness to new ideas
Processing	Focussing	Multidisciplinary awareness
Analysing	Exchanging knowledge	Coping with Ambiguity
Recombining	Presenting Knowledge	Team skills
Cooking / Processing	Relationship Building	Leadership
Presenting & Communicating	Networking	Learning from Zero

works is a model of mobility and communication across networks. And for this, completely new sets of skills are required.

We should not be surprised at our difficulties in applying these principles to our school systems overnight. The economy generally moves much faster than labour markets are prepared to follow, and educational systems are even slower to follow. Nevertheless this propensity to slowness is a threat where change moves so quickly.

No single answer

What does this mean for our education systems? It takes roughly 15 years for a student schooled in New Economy skills to go through the school system and enter the workforce. We are only now seeing the industrialised, Panopticon-schooled students entering the workforce from (then Singapore Education Minister) Dr Goh Keng Swee's educational reforms of 1978.

In August 2000 I was talking to some workshop supervisors from a heavy machinery manufacturing company in Singapore. I was discussing reactions to change in the workplace with them. These were men in their 40s. "Who provides the biggest obstacle to implementing change in your workplace?" I asked, thinking silently of the numerous studies that say that men over 40 are the most conservative elements of the workforce. "It's the youngsters in their teens and 20s," they said unequivocally. "Not the older workers?" I asked, somewhat surprised. "No," they said, "we have had to change all our working lives...even our apprenticeship was designed around moving us around different jobs and skills. We *like* change, we can learn more, we can get new ideas about how to solve problems. The youngsters don't see this. They don't like to do anything if they are unsure about how to do it."

Somewhat shocked at this turning of the tables on my preconceptions, I was immediately reminded of a question I used to ask Singaporean schoolchildren when they were applying to study English at the British Council. "What's your favourite subject," I used to ask, "Maths or English?" "Maths" came the reply nine times out of 10. "Why?" I used to ask, hoping for a new or different answer. "Because there is always an answer."

The trouble is, that in the new economy, there is not always one clear answer. Once the world is truly borderless and all the walls are down, the rules of the game change. The game is not now about capturing all the value in defined spaces. The game now is about movement through networks and webs. It is about mobility of capital, mobility of ideas and knowledge, and mobility of people. And there is no limit to that mobility. This is an environment of ambiguity rather than definition, an environment of confusion rather than clarity, an environment of adaptability rather than stasis.

Indeed we can see this in the business world, through the broken barriers between biology and technology in the burgeoning biotechnology discipline, in the convergence of the pharmaceutical industry with the personal care industry in cosmetic products, in the convergence between communications and computing, the convergence between technology, biology and physics in nanotechnology. The walls are gone. There is no single answer any more, and we have not been taught how to recognise that single answer. We have to try a variety, and see which ones work for us. Our graduating students this year are going to be very upset.

Ditching the panopticon

Dr Goh Keng Swee's educational reforms of 1978 were enormously successful in getting Singapore industrialised in record time. They streamed, packaged and distributed education in an extremely effective manner. Unfortunately, they were preparing students for an economy that was already in decline. Peter Drucker had already pointed to the rise of knowledge work as an economic force ten years before in 1968, and the convergence of telecommunications with personal computing exploded into networking power and the computer-driven new economy all through the 1990s.

If it takes 15 years for a school student to become trained in new economy skills, it takes 40 years to work an old economy teacher through the education system.

At the Asef Education Conference hosted in Luxembourg in May 2000, Dr Robert Downer of the University of Limerick stated that the biggest obstacle to educational reform lay in the educationists themselves. He pointed out that of the 60 European institutions that had survived intact and substantively unchanged from the Middle Ages, 56 of them were universities. That is a terrible testimony to conservatism and resistance to change.

> In 1968, Peter Drucker observed that there had been little change in tools and none in teaching methods in over 8,000 years of literate human history. "The only new teaching tool," he says, "has been the printed book. And that, few teachers really know how to use – or else they would not keep on lecturing what is already in the book." (*The Age of Discontinuity*, 1968, p.421).

Even now, with a rhetoric extolling the use of technology in education, great words about using IT to scale the costs of learning to ever wider classroom audiences, we are still using old economy panopticon models of boxing and packaging content and broadcasting it willy nilly to students who may or may not be paying attention. In an attempt to engage the deeper self through "interactivity", we use Skinnerian behavioural conditioning techniques that belong to the last gasps of the old economy in the 1950s. Students peck at the keyboards like Skinner's pigeons pecking at the correct button to get fed some grain, and lo and behold, when our students get the right answer, they get an animated fish and a canned audience cheer. (Clifford Stoll, *High Tech Heretic*, 1999, p.15-16). This is not *real* learning, not if pigeons and monkeys can do similar tricks.

Maximising strengths

Teachers proudly display internet-driven projects produced by their students, but privately worry that their students' mastery of cut and paste technology does not extend to giving an articulate oral presentation of their projects' contents. Cutting and pasting is not the same as gathering, cooking, digesting and processing knowledge, and then communicating this successfully to peers – and somewhere between our enthusiasm to show success in implementing e-learning among our students, and our anxiety about our own ability to cope with it as adults, we do not always check that the latter activity is also taking place.

And yet we must remind ourselves that we old economy trained teachers do have things that are of value to this hurried, hasty, flighty generation: we have experience, judgement, deep thinking and reasoning skills, abilities cultivated in a quieter age.

We can say with Master Confucius, "The little children of my school are ambitious and too hasty. They are accomplished and complete to a point, but they do not know how to restrict and shape themselves." (*Analects*)

It is here, in the humble awareness of our inadequacies, that we can begin to see a role for ourselves in enabling our successors to shape and direct their talents and instincts.

If our old economy teachers are still with us for the next forty years, we need to learn quickly how to minimise our old economy constraints and maximise our old economy strengths. We will shortly see how the role of performance management systems is going to become increasingly relevant in the hitherto closed and safe world of education.

6
Branching Out: Dilemmas in Education

So here we are, at the cusp of a new century, a new millennium, no less, in the midst of enormous change. We know that we are woefully inadequate in the face of the new challenges that face us, and we know that our education systems need to become more flexible and adaptive to help the next generation cope better than we can. We are miserably aware of the driving imperatives on the one hand, and our woeful inadequacies on the other. We know we are substantively blind when it comes to the new landscape, so we are timid about the extent to which we can be prescriptive – we declaim most loudly on generalities, and fall strangely silent when it comes to specifics. We know for sure, however, that our education systems need to become more flexible.

We do know, from our experience in flexible learning with *adults*, what seems to work well in flexible learning models. We know that adults embarking on flexible learning programmes have greater chances of success if:

- They are relatively young.
- They have had prior academic success in a formal educational setting.
- They have good time management skills.
- They have set very clearly defined goals.
- They have had pre-programme counselling and orientation.
- There is prior learner training and self-awareness training.
- They have had the opportunity to articulate and discuss their expectations.
- They have access to regular meta-counselling during the course on personal, life and self management issues.
- The course content and materials are composed for the flexible delivery format, and not adapted from traditional classroom based materials.

- There is a reliable and accessible support infrastructure.
- The programme provides opportunities for networking and peer support.
- There is a strong mentor relationship (internal or external to the programme).
- There is a variety of interaction opportunities, including regular face-to-face.
- The pace and sequence of delivery can be negotiated according to individual learning styles and preferences.
- They have a real world "laboratory" where they can try out and apply their new learning.
- There are clear, regular waystages with review, validation and feedback.

But we also know that flexible learning has only been extensively tried out with working adults, and relies heavily on the maturity, self management skills and self-motivation that work experience brings. Introducing flexible learning on a large scale to the pre-work education sector opens up a number of significant questions and dilemmas about the ability of both students and teachers to cope with these new approaches and models. The dilemmas become more specific when you look at the expectations for change over the next five years among Singapore educationists.

A group of educational professionals from the Institute of Technical Education, polytechnics and secondary schools, when asked in May 2000 for their five year predictions, saw further imminent change in four main areas:

1. Expansion in E-Learning
2. Expansion in Flexible Learning and increased collaboration between institutions
3. More autonomy in educational institutions
4. More "real world" contact and relationships

The implications for teachers were significant, they thought.

Predictions for education

1. Expansion in E-Learning imposes even greater demands on teachers in the area of new pedagogical literacies and new skills even beyond simple technical and operating skills. The movement towards project and internet-based teaching outcomes involves an increase in the teacher's ability to negotiate learning objectives with students. The unpredictability of the student's learning artefacts opens up enormous questions of comparison, assessment and moderation across a cohort, or in relation to the stated institutional teaching objectives. Flexibility also means potential for large divergences between learning outcomes re-opening questions of assessment and validation from another perspective. It also challenges traditional notions of syllabus and curriculum boundaries.

2. Flexible Learning, in addition to issues already raised, imposes increased flexibility and adaptability on the teacher. Increasing portions of the teacher's work will take place outside the classroom, and outside classroom hours. They will need to evolve a greater portfolio of roles, ranging from co-learner, to collaborator, facilitator and consultant.

3. Institutional Autonomy will bring divergent pressures in other ways, giving schools the freedom to evolve their own personalities and profiles, raising questions of interface with the national curriculum, and standardised measures of performance and ranking across the educational sector. Greater autonomy in teacher hiring, firing and promotion will bring issues of performance assessment and measurement to the fore for many professionals, and issues of survival and development will become much more prominent on the teacher's personal professional horizon. Opportunities for career mobility across the sector will be presented by professional development concerns, but challenged by divergent evolution of styles and standards between autonomous educational institutions. Performance management of teachers will become a much more significant function within schools.

4. Real World Contact will require teachers to build and maintain more external contacts, relationships, networks and experience, both with the community at large, and in the corporate world. Again, such imperatives will extend the teacher's traditional workspace beyond the time and space of the term-bound classroom.

The collective pressures of these changes will challenge the teaching profession in Singapore with five principal dilemmas:

I. The goal of student ownership and autonomy vs pedagogical control and direction

II. The likelihood of unpredicted outcomes vs the desire to achieve planned outcomes

III. The likelihood of novel outcomes vs the ability to measure and validate outcomes

IV. The likelihood of divergent outcomes vs the desire to coordinate and moderate performance across a cohort

V. The perceived need for negotiated outcomes vs the maturity levels of the students

Teachers will freely acknowledge that they were never trained and are poorly equipped for such Herculean endeavours. Nor could it have been anticipated that they should be. Again and again, we are presented with the compelling realisation that we know very well the *form* of the tasks that await us, but we are sadly inadequate as to the *substance*.

Does anybody really and truly understand the magnitude of the changes and transformations we will have to effect in ourselves to achieve a truly effective education system for the new economy? Probably not. These are new territories for everybody. Or are they?

Pepper

Picture a small country, hitherto lingering in relative obscurity at the edges of history, standing at the threshold of a new century. There is a new-found prosperity and sense of confidence in the air, based on growing success in trade and commerce. Luxury goods that were formerly only the possessions of the rich and powerful are increasingly within the achievable aspirations of the common citizen. There is enormous socio-economic change, with violent shifts in the employment market – new businesses operating under new rules, spawning entirely new industries, new skills, and new employment needs. Exciting new media are beginning to have significant impact on the communication and transmission of knowledge. In the midst of this, there is a vibrant debate about reform of the education system, emphasising a radical shift away from rote learning, content driven syllabus, conformist, traditionalist approaches to teaching. The radicals favour a more flexible, student-based, enquiry-oriented approach.

Time and place? If this sounds like Singapore in the year 2000, you would not be mistaken. But it is also a pretty accurate description of England in the year 1500, with sudden, dislocating shifts from a feudal, agrarian economy towards a mercantile bourgeois economy founded on the trade in luxury goods. The printing press has been in existence for only 50 years, but has already started to revolutionise mass communications and learning in Europe. And the education systems are coming under increasing pressure.

What drives this change, at the transition from the 1490s into 1500? Pepper.

The possession and conspicuous consumption of spices throughout the Middle Ages was a symbol of status, power and wealth, as well as a mysterious, potent symbol of Paradise, located in some unknown country far to the East, the fabled locale of the Christian Garden of Eden. Princes and Kings would make presents of spice to their fellow-rulers, along with gold and jewels. For a consumable and perishable commodity, it held a mystique and a value that we can today barely comprehend, until we remember our current obsession with our own strangely transient intangible, knowledge.

Europeans in the Middle Ages didn't know where spice came from. It was purchased for centuries in Asia by Arab middlemen and shipped to Syria and Egypt, from whence it entered Europe via Venice and other key trading centres. It was scarce, enormously expensive, and explosive in its impact on the European taste buds as well as the European imagination. The German historian Schivelbusch (*Tastes of Paradise*, 1993) describes the commodity pepper as the equivalent of petroleum in the 20th century, as the single most powerful economic driving factor in the European Middle Ages.

By the mid-15th century, two factors conspired to revolutionise the spice trade in Europe and revolutionise European economies and societies in the process.

First, a vastly increased demand. A rising middle class was acquiring the taste for pepper. Middle class recipe books from the period started using pepper sauce in quantities that even chilli-hardened Singaporeans would marvel at.

Second, a fast declining supply. The Mamelukes had come to power in Egypt, and the Turks controlled Asia Minor. Through a combination of blockades to the Christian West and exorbitant customs duties, the supply of spice through the traditional overland route was suddenly and rapidly drying up, just at the time when demand was mushrooming beyond control.

As a direct result, a long-stagnant transportation technology was stimulated into frantic activity by the realisation that an alternative sea-route to the fabled "Spice Islands" would bring enormous riches to those that discovered it. The Portuguese and Spanish mariners of the 15th and 16th centuries pushed back the boundaries of navigation and shipbuilding technology on the back of an insatiable European demand for pepper and the profits that they could bring, to be succeeded by the seafaring Dutch in the 17th century.

Whole industries sprang up to support this endeavour, from shipbuilding and supply, to military and naval infrastructures in support of economic interests overseas, to an explosion in subsidiary commerce and trade. European society started a radical shift from an agrarian to a commerce and craft model. New skills and educational methodologies were required, and the recently invented printing press provided the means and the technology to accelerate social change faster than had ever been possible in previous human history.

There are many obvious parallels with our own situation. But what is the significance of these parallels? Perhaps the greatest is the comfort we can draw in the midst of our bewilderment about what is going to hit us next. We have been here before. We may not know exactly how we are going to manage the discontinuous change that disrupts everything we do, nor how we can best equip ourselves to survive and prosper in a changing world. But we can find some reassurance in reminding ourselves that when circumstance and technology conspire, we are stimulated into a remarkable inventiveness and adaptability.

The outcomes of the pepper revolution in Europe started with the glorious canals of Venice, the first European city state to profit from spice, and culminated in the serene and civilised canals of Amsterdam, capital of a nation that fought its way from vassal state to Spain, to proud and global economic significance at the head of a world spice monopoly. Even the last vestiges of the old maritime pepper trade financed mercantile expansion in the newly created United States of America, and having prompted its initial discovery three hundred years before, proudly announced the graduation of this naive but adventurous newcomer on the world stage.

The civilised canals of Amsterdam may have been bought through barbaric and markedly uncivilised practices at the sources of spice in Asia, but the economic momentum this trade engendered also paid for the best achievements in European art, architecture, and humanitarian thought, as well as completely transforming the nature of global commerce, and revitalising markets as diverse as India, Korea, Formosa and the Chinese mainland.

In these confusing and tumultuous times, it is good to be able to reassure ourselves that change may also bring progress and opportunity beyond our imagining.

7
Sisyphus

We, as teachers, have a two-fold role with our younger students. We have the responsibility to impose and compel the skills and qualities that will enable our students to learn for themselves. To this extent our role is similar to the traditional mode of the master imposing knowledge on the student, whether willing or not. Simultaneously, however, we need to be progressively building the capacity in our students to generate and sustain these key learning skills for themselves. Our goal is ultimately to emancipate our students from our initial guiding control, into a relationship of collaborative and validating dialogue. This second task is the one most difficult to achieve, and the task too infrequently addressed.

Interacting processes

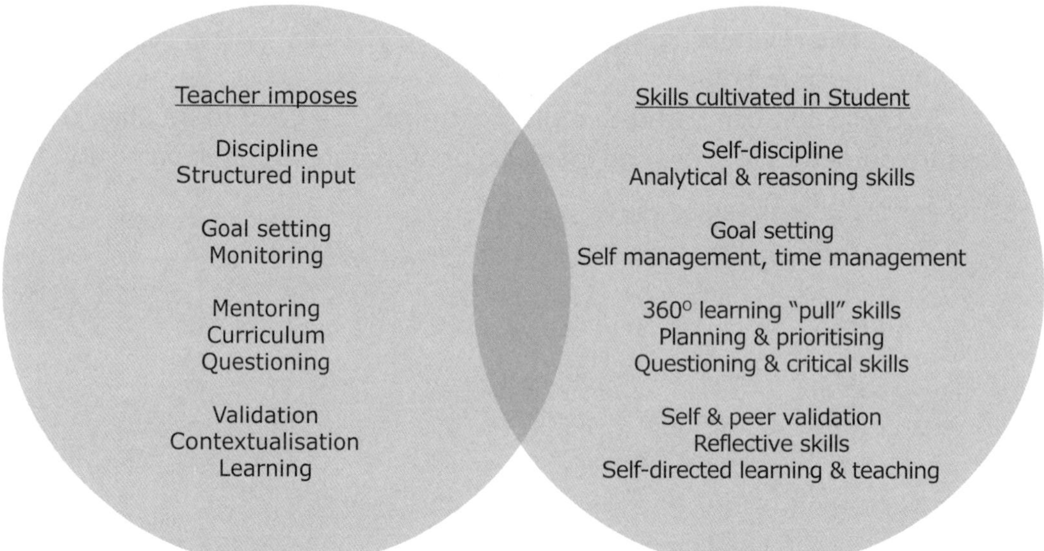

There is no doubt but that the pressures to change are immense. The scope of our choice in the matter is limited to whether we can seek to excel in the new environment, or retire to quiet and dusty backwaters where we shall be minimally disturbed.

Long-heralded changes

A noted guru of the knowledge economy put our position as educationists most succinctly by stating our three most central agendas. All of them reinforce our instincts towards more flexible models of learning.

1. We need to cease the squabbling between behaviourist approaches (stressing control, repetition, imposition of mental habits) and cognitivist approaches (stressing learning as understanding, self direction, significance of personal insight), as if only one of them were correct. They both reflect the reality of education, but from different perspectives. Cognitive learning cannot take place spontaneously without the external insertion of knowledge foundations upon which to build. The issue is not which approach is correct, but how to work the relationship between them to best effect.

2. We need to unfocus our obsession with different ability levels, and with designing systems that stream and address these notional abilities, because people generally have broadly similar innate abilities. Where we differ is in individual timetables, pace, rhythms, attention spans, and styles of learning. We need systems of learning that can be geared to individual learning profiles and accommodate them, while also providing for "choral", orchestrated learning contexts that also have significant learning impact for certain types of learning.

3. We need to remove from the teacher all the classroom management and babysitting duties imposed by the system that puts 30 to 40 kids into a box and takes them lockstep through a textbook drill for six hours a day. We need to liberate our teachers to truly teach. This is such a simple idea, but so dreadfully revolutionary – 90% of our training and experience is in babysitting, he says, and not in teaching.

He summarises thus: "Teaching and learning are bound to undergo tremendous change in the next few decades. Economic necessity forces us to tackle the job, no matter how great the resistance of citizens and educators."

These appear to be very apposite and timely words. Readers may not be very much surprised to know that the guru in question is Peter Drucker. Perhaps more disconcerting, however, is the realisation that he wrote these words in 1968, in his landmark book *The Age of Discontinuity*.

1968 in Singapore is a long time ago. The British Forces are still in Singapore, and independence from Malaysia is only three years old. Jurong industrial estate and new town is only just taking shape. The Employment Act is about to impose industrial age conformity and regulation to an unstable and rebellious labour force in Singapore. It will be another 10 years before the sweeping and dramatically effective educational reforms of then Deputy Prime Minister and Education Minister Dr Goh Keng Swee, designed to produce a world class industrialised economy workforce – highly structured and decidedly old economy in design and intent, and enormously influential in Singapore's economic success.

Déju vu

But there is a sting in the tail of our success in creating an industrialised economy: we have a generation of blind school teachers trained in a paradigm that doesn't work any more, and we are now seeing the full effect of those reforms emerging in full, focussed, panopticon strength into our workforce.

And here we are again, examining the very same issues that Peter Drucker described 34 years ago, or, for that matter, the same issues faced by the English humanist reformers at the dawn of the 16th century.

You will be familiar with the Greek myth of Sisyphus, who through pride and arrogance incurred the wrath of the gods, and was condemned forever to roll the same boulder up a hill again and again. When he reaches the top, the boulder rolls to the bottom, and he is compelled to begin the same weary journey all over again, achieving nothing.

There is a very compelling sense of *déja vu* about the issues that face us now. And indeed it makes sense that in the face of pressing change each generation must disassemble and reinvent its education and learning systems. It also makes sense that we are more poorly equipped than the generation we are attempting to teach. However, as with the blind tour guide, so long as we acknowledge our inadequacies, and know our strengths as well as our weaknesses, we can do a creditable job and gain the respect of the generation that follows us.

For centuries, Sisyphus was judged to be a doomed and unhappy man. It took the French philosopher Albert Camus to recognise that there might be joy in the struggle to achieve authenticity for our own generation at this fascinating moment in history.

> "The struggle towards the heights is enough to fill a man's heart. One must imagine Sisyphus happy."
>
> (Albert Camus, *The Myth of Sisyphus*, 1955).

8
The Geeks Have It

Who's going to survive better in the new economy? Subtle hints suggest that women, geeks and criminals might have a better chance than others.

Early warnings from regions that saw the collapse of old-economy industries and painful reconstruction of new economy businesses are giving tell-tale signals we should heed.

> In the 1980s, the coal mining industry in the UK collapsed almost completely. In 1945, the industry employed three quarters of a million workers – now there are around 9,000. A quarter of a million workers lost their jobs just in the last 15 years.

Woman power

Research from the University of Cardiff suggests that women from the traditionally strong mining communities have adapted much better and much faster than the men to the collapse of the mining industry. From previously housebound roles supporting a working man and growing family, they network, become active in community projects, and take new economy jobs in call centres, while the men go to the gym to prop up their redundant masculinity as defined by a tough physical life at the coal face. Steroid abuse is common. Women have become dominant both in the home and in the community.

In Sweden, the government is concerned at the number of women from the provinces who are migrating to the cities to find new economy jobs. The men prefer to stay behind in the mountains and forests where they can continue their hunting and fishing.

> "The men have hard, physical jobs, and they get away from that by going snow boarding and hunting with their buddies," explains Charlotte Svensson, political advisor to the Swedish Prime Minister's Office. "The men have these networks where they are accepted; they have nothing to gain by going to the cities. The women, on the other hand, have nothing back at home except the household chores. The cities offer them many more opportunities."

The new economy demands skills in dealing well with uncertainty, ambiguity, and unreliability, says Professor Bob Fryer of the University of Southampton, and it requires adeptness in solving problems, and creative, speedy solutions using the materials to hand. "Women are much better at all of this because they have been doing this all their lives – after all, they live with men."

At least they used to. The new dominance of women in these societies is creating worrying social trends. In the former mining communities of Yorkshire and South Wales, women are having children by their boyfriends, but refusing to let the fathers live with the families.

Social psychologists and evolutionary biologists appear to confirm this perception: it makes sense that women function better in networked communities of tacit knowledge exchange. While men's brain structures are much better at spatial and physical perceptions, women's brains are much better at processing and analysing emotional and attitudinal information from the people around them.

The men might strike back by pointing to greater decisiveness and aggressive goal-setting activities – "It's all very well exchanging all this information, but you've got to do something with it" – but the fact is that muscles, spatial skills and physical prowess matter little in cyberspace.

> Silicon Valley Knowledge Management expert Dan Khanna on a trip to Singapore, joked that Southern California was full of golden boy football hero hunks earning peanuts in hotel carpool jobs, while the weedy nerds had made billions tapping maniacally at computer keyboards.

The criminal element

And where do the criminals come in? Criminals are by definition highly skilled and experienced in the art of innovation. They have to innovate in order to stay ahead of the systems they are attempting to subvert. Law, security and bureaucracy always follow the innovations of "miscreants"– this word itself acknowledges the creativity intrinsic to the role, while damning the direction this creativity takes. Two things give added opportunity to the criminal innovator: transition, and technology.

The turn of the year from one season to another is expressed in cultures worldwide as the opportunity for the underworld to slip back into our world, slipping between the cracks in the seasons, to threaten us with other-worldly desires and motives. While the Chinese celebrate Hungry Ghosts Festival and buy the spirits of the long-dead off with food and entertainment, westerners anticipate Hallowe'en, with its paraphernalia of trick or treat, bonfires and lanterns to lighten the threatening night, and fireworks to scare evil away.

Auditors will tell you in eerily similar tones that transition from one business system to another creates the greatest opportunity for criminal fraud, and it is their difficult job to imagine the directions the underworld mind will take in reaping grim advantage from the opportunities created by change. In speaking of transparency, consistency and inspection, they too light their lanterns and make loud noises to keep the underworld at bay. Accountants hate and loathe change for similar reasons – and they will slow it down if they can, not out of mischief, but out of prudence.

> In the 1990s, the Adriatic Sea suffered a strange phenomenon in the summer. Over the course of just a few days, a red plankton would turn the sea from Rimini to Split into a sea of blood. The build up of fertilisers from farming and nutrients from industrial effluent pour into the sea from the Po valley throughout the year, but it is the first hot days of summer sunlight that cause the plankton to bloom suddenly and violently, sucking the oxygen out of the water and killing tens of thousands of fish.

In the same way, the conditions for rapid and constant change have been building in industrial societies for decades. The convergence of telecommunications with computing technology created the conditions for the violent and rapid change that now rips and eddies through every dimension of our society. In many ways this brings enormous opportunity; but it also creates conditions for a feeding frenzy for black economies and criminal elements, from money laundering to human trafficking, from hacking to cyber-blackmail to viral network bombs. We know we have only seen the start of what is possible, and that some high profile events, such as the attacks on Yahoo and Microsoft's Hotmail, are really only experimental preludes to more serious subversions. Meanwhile, legislation struggles to keep pace.

Technology too provides enormous opportunity. Technology scales knowledge out of proportion to its physical location, and out of proportion to social or economic position.

> In May 2000, a computer school dropout in Manila unleashed a virus that crashed networks and computers from the Pentagon and UK Parliament, to corporations like Ford and Microsoft, causing billions of dollars worth of damage. Filipinos hurried to condemn the act, and echoed the outrage of the powerful, but they also found it hard to resist a sense of celebrity. Anyone can have a global impact in the new economy, and the culprit, 24 year old Onel de Guzman was reportedly being wooed by computer security companies.
>
> Yet the same society two months later was reminded of the tragedy of its real economy, when the Payatas rubbish dump on the outskirts of Manila moved, and buried between 180 to 1,000 scavengers alive.

Political and social theorists worry about the scalability of knowledge in terrorist or criminal hands – the knowledge that leads to bio-genetic warfare, nuclear bombs, sarin gas, is disturbingly accessible and easy to implement. And criminals have always used technological innovation to stay ahead of the law.

Crime and innovation

South American drug cartels employ first rate chemists and run state of the art labs; pornographers have pioneered new media and distribution channels for centuries. It is often supposed, for example, that it was the underground market for pornography that made the VHS videotape system the global standard, and brought VCR and videotapes into the realm of consumer affordability. Even now, DVDs are still slugging it out with VCDs on a battleline drawn up between a technology supported by big business and a more accessible reproduction technology utilised on a vast scale by pirates and pornographers.

Organised criminals are entrepreneurs. They too are out to make a profit through innovation. The main difference between these entrepreneurs and (most) legitimate business entrepreneurs is that they don't care about the destructive social impact of their actions. But they have an enormous advantage: they are not constrained by heritage systems and ideas. Simply by being illegitimate, they are liberated to innovate in extraordinary ways.

It follows from all this that societies and organisations that are able to harness criminal innovation, the new economy networking and collaborative skills of women, and the solution-hungry intellectual frenzy of geeks, will do better than the societies and organisations that do not. This is neither a flippant nor an irresponsible claim; and I do not mean to suggest that criminals be encouraged or that everybody be trained in feminine or geeky qualities. What I am suggesting is that qualities that enable success in the new economy often come from outside traditional and legitimate structures of power and influence, and the more we can learn from those disempowered, illegitimate or ridiculed sources, the better we will be able to position and protect our societies in an unstable and fractured world.

In my next chapter I shall explore Singapore's performance and potential in learning and acquiring insurgent new economy skills.

9
The Real Threat to Asian Companies

In my previous chapter, I suggested that we have much to learn from women, geeks and criminals in the new economy. Women are better than men at networking and collaboration, geeks pursue solutions with single-minded energy, and criminals innovate constantly.

How does Singapore perform against this rather unconventional matrix? It's a rather mixed scorecard. Singapore is an extremely hostile environment for criminals, and by international standards, sees little criminal innovation in any organised way. It is often portrayed as a nation of single-minded geeks, however, and it is true that there is a very pragmatic, solution-driven approach to many aspects of life here. With 53% participation in the labour force (and growing) Singapore is also a society of apparently empowered women

It is often noted that compulsory national service for men gives women two years' head start in the market for jobs and experience, and women do land senior jobs with greater regularity than in many other countries.

> Kin San, a secondary school teacher in a local school disagrees, however, that working culture is feminised in new economy directions as a result. "National Service is an exercise in social engineering as well as one of security and defence" he says. "It brings together people of all races, classes and backgrounds, and teaches them how to work together. But it also imposes a command and control view of the world, which the men bring into the workforce with them. The women follow – if they want to get on in the labour force, they have to follow the same old command and control system."

Collaborative networks

In a recent interview, I was asked if Asia had any special challenges in adopting knowledge management practices from the West. In thinking about my answer, I realised that Asian cultures actually have quite a strong advantage over the West. Asian cultures are generally very good at developing, growing and exploiting collaborative networks, within which knowledge exchange and movement is extremely efficient. In the United States, by contrast, three hundred years of a gospel of competitive individualism inhibits collaborative knowledge exchange among peers.

In an Asian company, there are often extremely highly developed lateral informal social networks of support and collaboration, which are rich in potential for knowledge exploitation. The horizontal social capital potential is extremely high. People are also much better adjusted to learning socially in Asia: everywhere you turn, you will see groups of young people working on school projects together. What's more, technology in general, and email in particular, amplifies horizontal communication and relationship building much more than it enhances the vertical information systems it was designed to serve.

In stark contrast, I vividly recall (an admittedly extreme) case of student paranoia from an American co-student at University. He refused to take his turn to present a paper at a study group seminar: "The b*****ds will steal my ideas," he stated, "Nobody's going to see them till I have them published in a journal."

In the United States it is much harder to cultivate and acquire such social capital potential, as commentators are just beginning to realise. American workers collaborate much better with their bosses (who can give them a promotion) than with their peers, with whom they are often in a tense, competitive relationship.

The real challenge to Asian organisations is in creating strong patterns of vertical collaboration and knowledge exchange – fairly rigid systems of hierarchy, power and decision making inhibit anything other than highly structured reporting exchanges. Last week a colleague of mine commented on the frustration that many Singaporean managers feel in relation to this: "How can we get our workers to communicate freely with us and give us feedback about problems and opportunities?" they ask.

> Twenty-seven year old executive Edwin Tham puts the case from the worker's point of view. He points to the command and control mentality as the chief inhibitor. "Managers keep speaking the propaganda of team work, collaboration and feedback," he says, "but *they* don't collaborate with *us*, and they don't involve us in the decisions." When Edwin joined his department, there was no manager in post, and one of the senior executives was acting head. "We were all working

very closely together," says Edwin, "and we shared the problem-solving among us. We knew when we had to call in someone more senior to help out. The work was very fulfilling and we all felt we had a stake in it."

Things changed when a new departmental manager was appointed. "The new manager made each of us report directly to him," says Edwin. Suddenly the collaborative relationships within the department were short-circuited, as the manager assigned tasks individually to the executives, straight down the line of control. Worse, tasks were assigned with no context or explanation, and executives were no longer involved in thinking about the decisions that needed to be made. "That was the final straw," says Edwin, "I decided to quit."

Now vertical collaboration knowledge transfer is critical for any organisation's performance, because it is only through such channels that strategy gets turned into action. American corporations are often much stronger and faster than competitors precisely because of their ability to turn strategy into reality – and the gospel of individualism rewards American workers for co-operating with their bosses more often than they co-operate with their peers. The Asian superiority-based command and control culture has also been extremely effective in turning strategy into action – while strategy depended on simple, reproducible ideas, that is.

The innovation reservoir

The weak point for American enterprises and for Asian command and control cultures is that in the knowledge business, speed of strategy implementation is not enough: you must also have innovation and creativity. Organisations that can exploit horizontal knowledge transfer and collaboration can multiply the innovation quotient of those organisations exponentially in relation to the numbers of people involved.

That is the hidden advantage of Asian, and Singaporean, organisations. If Asian managers can release this horizontal social capital and harness it in support of strategy implementation, they will have enormous innovative potential at the service of their enterprise. Entrepreneurial Singaporean startups are already learning a lot about such skills – I suspect Sim Wong Hoo's Creative Technology could teach a great deal about the exploitation of horizontal social capital potential. Where the obstacle for Americans is lateral communication networks, the obstacle for Singaporeans is vertical communication networks. In both cases, whether it is strength in vertical collaboration or horizontal collaboration, the critical issue is one of trust; and so the battles for sustainable success in the new economy are going to be fought on those decidedly un-technological fields of trust, influence and collaborative leadership.

There are already interesting signals that horizontal networks are beginning to subvert even the traditional boundaries of companies and organisations. Think about it: email knows no boundaries. You can communicate with your colleague next door just as easily as with a business contact in another corporation in Japan, whom you happen to get on with very well.

E-markets

> The buzz word for year 2000 business innovation was e-markets. For example, in that year Forrester Research predicted that e-markets would represent 75% of all B2B online transactions by the year 2010 – meaning roughly US$21 *trillion* annual turnover within a decade. We already know about the Ford AutoXchange, and the GlobalNetXchange, which in 2000, represented US$200 billion worth of business each for the automotive and retail industries respectively.

Other electronic markets, hubs and exchanges, followed hard on their heels, in catering, transportation and logistics, fiber and wood products, and many, many more. Not all of them succeeded, but you can see the powerful arguments involved for big corporations: if you can automate your entire supply and value chain, if you have real time knowledge transfer between participants in the market, you can remove billions of dollars of inventory from your warehouses, and put it on the road, and you can shop for the best prices instantaneously.

But what will e-markets do to companies? What you are doing in an e-market is effectively hardwiring your back office ERP systems into a global marketplace. Everybody in the marketplace will have access to the same technology and the same information about products, prices and customers.

Think about it: competition in the non-automated marketplace is now primarily an array of "arms races" in technology, pricing/costs, war for talent, and market knowledge. The net market automates everything except the war for talent. Suddenly, everybody is absolutely equal in technology, has equivalent access to price/cost structures, and customer knowledge. It is therefore only on the field of talent that future competition will focus – and competitive advantage through innovation is synonymous with the war for human and social capital.

The net-markets also connect communities of practitioners across companies in the cyber-marketspace. These people will be communicating more frequently with their peers across companies than with their line managers in their own companies, because that is what the net-market does best. It turns your company inside out, so that it is completely customer and supplier focussed – in sharp contrast to the old economy conglomerate, which spends the majority of its effort on internal systems and control.

> We already see tangible representatives of this cyber-realm in what Kenichi Ohmae calls –"virtual single companies" – symbiotic networked companies like Cisco Systems, Nike, Gateway or Dell.

In such a situation, it makes much more sense for me to evolve cross-marketspace relationships with my peers and collaborators in other companies within my network, because these are the people I work with every day. Increasingly, my line managers are going to have less impact on decisions affecting my work: the technology will make some of them, and I will make the rest with my peers in my cross-market community of practice.

> The British humourist writer Douglas Adams, author of *The Hitch-hiker's Guide to the Galaxy*, imagined his characters slipping back through time to just after the era of the dinosaurs. There, they discovered that mankind had evolved from a spaceship-load of middlemen and middle managers marooned on earth millennia ago. The planet they came from had found that their middle managers were slowly strangling creativity and innovation across the planet, and they resolved to be rid of them. So they convinced the middle managers that an asteroid was about to hit their planet, and that only the best minds should be saved from total destruction – of course, that would have to be the middle managers. They were duly packed into a spaceship and sent into deep space. The home planet suffered a dismal fate as a result, but Earth? Well, that's a different story.

As a middle manager myself, I do not mean by this anecdote to demean this rare, crucial and threatened breed. But it speaks truth at an important level; middle managers do indeed have the capacity to destroy innovation, communication, trust and collaboration, and many of us do so routinely. Edwin Tham's manager certainly did, as we saw earlier.

The insurgent factor

I would go further, and suggest with Harvard guru Gary Hamel that the greatest threat in the new economy is a generation of conformist, command and control leaders who do not know how to challenge norms, ask questions, and who do not seek to learn from unconventional, lower status, insurgent sources.

Singapore has a reservoir of geeks it can call upon, but it constrains their energies in conventional directions. It has women in positions of influence in the workforce, but does not especially exploit new economy skills associated with feminine characteristics. It has immensely hard-working, ethical and committed workers, and has the advantage of strong lateral collaboration networks, but it has not yet worked out how to harness this immense social capital strength to inform and implement strategy. It is insulated from innovation because until very recently it has discouraged insurgents.

The good news is that change is upon us whether we are prepared for it or not. Horizontal communities of practice that do not respect national, company or organisational boundaries are already becoming social norms, and they are already spreading innovation faster than traditional hierarchical structures can.

The sooner we, the middle managers, school ourselves in collaboration and trust, the sooner we will be able to create advantage from this for ourselves, our families, and our country.

10
In Praise of Older Workers

In early 2000, I observed two "new economy" events in Singapore where feelings ran high and frustrations were barely concealed. The first was a Forum on Small and Medium Enterprises, a briefing on the Singapore Productivity and Standards Board's new report, SME21. The second was the 23 April 2000 *Talking Point* show, discussing the Ministry of Manpower's new Singapore Learning Initiative. At both events, buzzwords buzzed, and officials were articulate, compassionate and understanding – but absolutely clear in their own minds that their audience had a mindset that needed to be changed.

So why was the audience exasperated? Let's take the *Talking Point* programme first: caller after caller pointed the finger at the employers who discriminated against older workers, when it came to jobs and when it came to training. Why should we go for skills upgrading, the callers asked, when the employers ignore it and give jobs and opportunities to younger workers? "Most employers in Singapore only pay lipservice to upgrading," said one caller. "They don't want you to upgrade because if you do, you will go for a better job." This caller knew exactly who he was talking about, and it wasn't the MNCs, whom he said were quite open to skills upgrading. Who then are these evil employers? Could it be the SMEs? If so, then we are in serious trouble, because they account for half the Singapore workforce.

SMEs and success

The SMEs had themselves been lectured only a few days earlier on another topic: upgrading and professionalising themselves for the new economy. At the SME Forum, hosted by the Singapore Institute of Management, the audience was made up largely of small and medium business owner-operators.

The message was an excellent one, showed deep understanding of the strategic difficulties facing entrepreneurs, and outlined the support measures being implemented as a result of the SME21 report. The audience did not cheer; indeed the man next to me could hardly contain his impatience, alternating between snorting at the speakers and checking his stock prices. It was the entrepreneur members of the SME21 sub-committees who compelled most. Occasionally the rhetoric slipped and reality seeped through.

Catherine Lam of Fabristeel looked us in the eye and told us that the Knowledge Based Economy could wipe SMEs out. We shifted uncomfortably – intuitively, we knew this was what lay behind the air of tension in the room. Nobody was having an easy time. We didn't need SME21 to tell us that most SMEs are isolated in a vicious spiral of crisis, firefighting, cashflow worries and the desperate search to contain costs and create enough excess margin to break into the next step of growth. Most people present didn't like being exhorted to break out of that spiral, to have trust in partners outside their family and associates, to cooperate, depend on others, to seek out support and grants for more professional management. We just wanted to know how to succeed in a merciless market driven by costs and uncertainties.

The sense of frustration in both cases struck me as being very similar. It was the sense of having your back against a wall, of being told you have to do better, of being given a list of suggestions about what to do, and of not seeing the connection between this list of (very) good advice and your next meal or your next contract.

It struck me then that maybe we are preaching too much, maybe we are describing the problems in too much detail, and maybe we are pushing the new economy rhetoric too hard. It's clear the need to change is urgent, it is clear that survival is at stake, and it's clear there is complacency among both workers and SME employers. The "no time no money" syndrome hits both – when times are good, we have no time to train because we are so busy; when times are bad, we have no money to train, we have to work to survive.

Success stories

But we are perhaps ignoring the success stories in our rush to instil a sense of urgency, and hence we are losing a sense of the possible.

> My mother left a small country school in Ireland over 50 years ago. She was 14 years old, and she believed she was stupid. She and her brothers had endured nine years of bullying, sarcasm and brutal discipline at the hands of a husband

and wife teaching couple. The husband was hated so much that his gravestone is still regularly toppled over in the local graveyard. He has been dead 40 years.

My mother went on to learn a trade – hairdressing – married, moved to England and set up a new home in a foreign country. She raised five children, three of whom are Oxbridge graduates, the other two becoming a nurse and a teacher (!) respectively. She must have sometimes wondered where all the talent had come from, but she never for a moment suspected she had anything to do with it. Last year, my mother completed her autobiography, a lucid and well-written series of reflections on her childhood, youth and married life. In many ways, this autobiography was her final, dignified answer to those ignorant, destructive teachers who had demeaned her intelligence in every way. It took her 50 years.

Whole generations of Singaporeans schooled before Mr Goh Keng Swee's educational reforms of 1978 were subjected to similar brainwashing. I do not for a moment suggest that Singapore teachers before 1978 were bullies of the degrading nature that my mother experienced. But society has systematically ranked workers based on academic merit, excluded them on the basis of poverty as well as academic aptitude from the possibility of upgrading, and given them a deep and lifelong sense of inadequacy. The classroom is the potent symbol of that exclusion.

Mary Phang (not her real name) is an all too common story in the HDB heartlands. She left school at the same age as my mother. Unlike my mother, she loved school, but her family had to make ends meet, so she left to get a job. She is feisty, smart, and rich in lifeskills; when it comes to calculating compound interest and probability theory she will hesitate, but her mathematical prowess knows no bounds when it comes to the chances of winning on 4D or Toto.

Her knowledge of HDB and CPF rules and regulations and how to navigate them to best advantage would leave an Oxford professor reeling. When it comes to advanced strategy, she is unrivalled at the mahjong table. Her extended social network gets her discounts, opportunities, jobs, and a life full of richness. She owns her own home, has managed a small business, has put her son through university, and although only five years older than me, is probably worth three times as much.

When it comes to lifelong learning, however, Mary's response echoes the third of the workforce in a recent Ministry of Manpower survey, who said that they were "too old" for that. "Headache!" is her response, and "My brain cannot cope." She simply does not associate her ability to learn in the school of hard knocks with the ability to learn in a structured way.

Even an enlightened learning organisation faces barriers like this:

> In 1997 the supermarket chain Cold Storage in Singapore launched an innovative learning initiative, called *Choose Your Own Training* – offering free upgrading courses from a selected list for any member of staff. A year later, nobody had taken it up. When managers asked staff about this, two replies came back: "We don't want to waste the company's money" and "We appreciate being given the chance, but it's like an MC (medical leave certificate) isn't it, and we don't want to put pressure on our colleagues by taking time away from the store."
>
> When Singapore Arlines Group launched its Communication Skills Programme (CSP) with the British Council in 1990 for mid-career workers with low educational levels, the programme faced similar difficulties, from both workers and their supervisors. Early participants faced scorn from their workmates for "going back to school", and some supervisors couldn't see the point of releasing their staff from a busy workplace. "Cannot Speak Properly" became the unofficial and unhelpful moniker for the CSP.

Ai pia jia ay ngeah, goes the popular Hokkien song --"You must fight before you can succeed, sometimes you are up and sometimes you are down, but you must go and struggle." It could be an anthem for the Singapore worker and small businessman. Both Cold Storage and SIA Group struggled, and overcame these barriers.

> In 1999, Cold Storage implemented the PSB's Critical Enabling Skills Training programme (CREST) on a massive scale – all seven modules, for all employees. It is an impressive move in an industry with razor thin margins, and an average of four months' job tenure in front line staff. Module One of CREST is called *Learning to Learn*, and asks participants to examine their attitudes to learning, find their own preferred styles of learning, and to seek opportunities for learning from a variety of different sources.
>
> "We saw immediate changes in our workforce," reported Diana Ong, Customer Service Manager for Cold Storage, at a recent conference on Intellectual Capital. "Suddenly our staff were coming to us to look for more training, and taking charge of their own development. Our *Choose Your Own Training* scheme suddenly came to life."

The confidence factor

Like Cold Storage, SIA stuck with their employees and took a comprehensive and systematic approach to show their seriousness. Coordinators and supervisors were briefed on the programme, and many of them were trained as mentors, to support the learning activity back in the workplace. The Chairman of the Group attended all graduations. A newsletter was launched to celebrate the successes of the programme, and social activities such as Scrabble clubs arranged. Graduates of each module attend a formal graduation ceremony, receive certificates and book tokens from a senior officer, and they make graduation speeches in front of their colleagues. It is deeply moving to see workers who believed they could never improve their English, stand in front of 200 colleagues and make a speech.

Gradually, support for the programme grew, success stories of CSP graduates who went for further training and got promotions began to spread, and people started asking to be nominated for the programme. In the past 10 years over 3,600 employee places have been taken on this programme, and almost 900 employees have reached an N'level standard of English. Forty six out of that 900 workers started from being unable to read or write in English at all.

One theme for the last 10 years of the programme comes through time and again in these speeches. It is not especially a theme of prowess or fluency in English. The theme is one of surprise at their discovery that they could achieve something they thought was already beyond them. The theme is *confidence*.

Our rhetoric is a little too harsh and it is a little too negative.

In 1983, the year the BEST English programme was launched in Singapore, 76% of the workforce had not progressed beyond secondary education. By 1998, that figure had gone down to 66% - slow progress, to be sure, but one that has ensured Singapore's current competitiveness. In the past seven years, 300,000 workers have been retrained through the PSB's innovative OJT programmes, stretching the concepts of learning away from the traditional (and feared) classroom. In the next five years, the PSB targets 500,000 more. In 1995, a *Straits Times* survey found that only 10% of the workforce had gone for formal learning and study opportunities. In 2000, a Ministry of Manpower survey found that 43% of those interviewed had undertaken such formal learning within the past three years. It was a busy five years.

The reality of the Singapore workforce and our SMEs is this: we are rich in talented, smart individuals who have managed resources, homes, small businesses, multiple jobs and families. We are rich in an immensely strong and proud people, who have struggled all their lives and built astounding success for their country. Some have broken out of poverty and some have not. They have as much to teach as they have to learn, and in our rush to preach the rules of the new economy, we forget that they know all about survival and learning to survive. It is perhaps time to sell our brand of learning a little better, and show our people the art of the possible. We need to be able to recognise non-academic experience and skills, and reward them, and the Productivity and Standards Board's National Skills Recognition programme will probably do just that.

Finally, perhaps, we can all agree that learning does not have to be a nasty or intimidating experience in a classroom. It doesn't have to take people away from their jobs. It can enrich the lives of employers and employees in financial as well as human terms. As employers and as employees we need to look to our strengths and our experience. Our older workers have a lot to teach us on that score.

11
Time and Motion

Watching workers work, in order to change how they worked, was a peculiar product of the scientific management movement of the early twentieth century.

> Almost a hundred years ago, Frank and Lillian Gilbreth set up cameras at the New England Butt Company in the USA, and began a six year photographic study of the physical movements required of labourers in completing tasks. (Frederick Taylor, the strict Victorian "father" of scientific management, had already been observing workers with a notepad and stopwatch for over ten years.) By 1914, the Gilbreth's shop floor was as wired as a Hollywood set, with some studies conducted in the dark, and workers trimmed with fairy lights to identify the precise motions of their limbs.
>
> Gilbreth's workers didn't seem to mind too much – they were, after all, the stars of the show, and many of his conclusions, while improving efficiency, also improved worker ergonomics and lessened their fatigue and stress.
>
> Taylor's approach was harsher, and he habitually fired workers who refused to do things his way. In his studies of the limits of worker endurance and productivity at Bethlehem Steel in 1899, he was said to have been so hated that he needed bodyguards to accompany him home from work.

This mechanisation, fragmentation and de-skilling of human labour into its component parts through observation and dissection now appears a distant legacy of a grimier, old economy, industrial age. Or so we might assume. Think again. The technology may have changed, but it is still being used to watch, monitor and measure the actions of employees.

The watchers

A survey of American employers in 1998 found that 20% of employers store and read employees' emails, and there is evidence that this is growing. About 54% of companies routinely monitor internet usage. 16% of companies routinely videotape their employees. There are now "active badges" that can track your movements through a building, and computer surveillance software that records desktop computer activity, keystrokes and down time.

Call centres and telephone exchanges are among the most monitored, and the most time-driven. Software logs you in, and tracks your call quotas, giving instant reporting to your supervisor. It logs you out, and records your breaks. In a Manhattan directory assistance call centre, the best operators can manage an average of one call every sixteen seconds. This centre manages 5.5 million calls per day. No wonder then that the word "Please" has long been shaved off the initial question: "Directory assistance, what listing?" In call centres, as in the 100m sprint, shaving a millisecond here or there can make a big difference. No wonder that telephone-related jobs have the highest turnover, and the greatest stress. At AT&T, where computer monitoring is standard, a quarter of the workforce is undergoing counselling for work-related stress (*www.robmagazine.com*, February 1999). And the more that computers, technology and integrated networks enter our working lives, the more we can be watched.

Watching and measuring is one thing. When the technology starts moulding us in its own image we should pay closer attention. Frederick Taylor imposed rigidities of control for his labourers via step by step instruction cards analysed down to the tiniest movements. They left nothing to chance and squeezed the maximum productivity from the working day. Divergence from these instructions while Taylor was on the factory floor meant instant dismissal. Palm Pilot enthusiasts rarely mention that they have had to change the way they write their alphabet, to get the machine to recognise what they write. When a "T" is written like a "7" and "A" is written like an upside down "V", you begin to realise that if this appeared on paper, we would be developing a new set of hieroglyphics pretty fast. Who rules the knowledge-based roost now? Man or machine?

People make the difference

On the larger scale, Enterprise Resource Planning (ERP) systems are famous for their cost, their complexity and their numerous failures. Less obvious but nonetheless significant, ERP implementations often face severe change management problems because they are so inflexible in relation to legacy IT systems, work cultures or work systems. ERP implementation is a massively disruptive exercise in information standardisation, and it forces an organisation not merely to standardise within itself, but also in relation to all its competitors.

ERP systems are so huge and complex, and their creators so resistant to adaptation (for how then would you sell the upgrade?) that they compel the organisation to re-engineer itself in order to fit within the processes described by the ERP modules. Making the wrong choice of software could mean an organisation redefining itself in a way that is incompatible with its strategy, goals or underlying culture.

> Dell Computers abandoned its ERP implementation when it found that it was incompatible with its new decentralised management model, and Applied Materials gave up when it couldn't cope with the organisational change issues the ERP system generated.
>
> Gartner Group reported in 1997 that fully 20% of all ERP implementations ended up on the scrap heap.

So not only are workers being technologized into robotic compliance, but enterprises are also being fashioned and sculpted into standardised, interoperable and constrained shapes. Globalization is one of the reasons for this. Interoperability and the growing importance of standards for data exchange between enterprises are a direct result of the networked economy. The more you are connected and the more you collaborate, the less idiosyncratic your infrastructure can be.

But this leaves a few big questions unanswered. On what basis do enterprises compete now? Gone are the days when they could compete on better processes (ERP levels all that) or better technology (heavens, we all have the same!). Very shortly we won't even be able to compete by getting an edge on customer or market intelligence. Access to information, standardised CRM and business intelligence systems (yes, the old ERP vendors are selling these modules too) is rapidly standardising across the global marketplace. Your competitor gets the same data, the same way.

Pretty soon (and in our hearts we know this already) the only competitive differentiator we will have left is people. Human talent. Given broadly the same technology, broadly the same information, and broadly the same business processes, only the people make the difference. And within that, the ability to innovate successfully is critical

– and by that I don't mean people who can have bright ideas. I mean people who can collaborate rapidly and effectively to turn existing knowledge and new inventions into value creating products and services.

The compliance trap

So we have to question the technology-assisted surveillance and compliance trends of recent years. Turning our people into compliant robots will only ensure that we all standardise to the technical landscape of wires and buildings and grey suits. Innovation will be dead, talent will be dead, and competition will be a meaningless word. Of course, that won't happen. Because there are still small enterprises out there (even in these troubling economic times) who do not and will not buy ERP or CRM systems, do not and will not watch their employees through video cameras and surveillance software, and who believe above all that difference, creatively handled conflict, and challenging the rules of the game will give them a competitive edge. The big enterprises who fail to internalise this spirit of non-compliance will become strangers to competition and like the crumbling Cold War bureaucracies of Eastern Europe, go the way of the dinosaurs, out-competed by our tiny, mammalian ancestors.

Technology and standardisation can have two effects on us. They can box and constrain us, compelling us to read and respond to the world in routines designed by unimaginative technocrats. Or, like a box turned upside down it can provide us a new platform to display our creativity at new heights.

The creativity guru John Kao uses the analogy of music. Music is bounded by rules and procedures, formulae and tightly bound relationships. Yet in jazz, the creative musician uses those rules and relationships as a framework from which to generate new experimental forms, working outwards from the rules rather than re-describing them. In the same way, language itself works to a structured grammar that describes how verbs and nouns and sentences should work. Yet in skilled hands, these rules (and sometimes their fracture) create new and astonishing forms.

It's time to stop watching our workers for compliance. It's time to start looking for workers who do things in new and interesting ways.

12
Gutenberg's Periscope

Most knowledge management theorists will rally around the motto: "If only we knew what we knew." In fact, quite often, it's not the knowledge that's resident within a company that's important. It's the knowledge that the company doesn't have, and doesn't even know exists.

> In the late 1430s, a German goldsmith, inventor and businessman wannabe named Johann Gutenberg, secured venture capital funding for a get-rich-quick enterprise. Every 20 years, there was a major religious festival and fair at Aix-la-Chapelle. Pilgrims flocked from all over Europe for this event. Gutenberg's invention was a periscope, which would enable the pilgrims to see over the tops of the crowd and view the religious processions and accompanying festivities. It was a great idea, and it was a great invention, but Gutenberg got the year wrong. His periscopes were ready a year early, found no ready market, and his investors took him to court and bankrupted him. Gutenberg got his context wrong.
>
> Gutenberg surfaces again 10 years later in Mainz, now in partnership with a businessman called Johann Fust. This time the partnership is working on four convergent technologies. First, Gutenberg is using his goldsmith skills to create moulds for creating typefaces; he is using winepress technology to arrange the type forms in a frame, and screw them down to create an impression in paper; he is developing a new type of fatty based ink, which will stick to metal and create a clear and durable impression on paper. One of his co-investors owns a paper mill, a technology only recently imported into Europe. He is the inventor of the modern printing press.
>
> This time, his context was right – not, initially because there was a large market for books, because there was not. The Christian Europe of that time was terrified of hell. Gutenberg's invention found immediate and lucrative markets in the printing of indulgences, pieces of paper issued by priests giving them time off their suffering in the underworld, in exchange for donations to the church. Gutenberg never made any money out of books.

Leaping forward 400 years, the fax machine is another good case in point. A fully-functional fax machine was invented around 1851, and Napoleon III was sending regular faxes between Paris and Lyons in the late 1860s. But the context was not right for commercial success, the fax machine was relegated to a minor role in the US Map Service, and it took 130 years and widespread commercial telephony before the invention found its place in history.

Risk takers and visionaries

Market readiness, and convergent technologies: there are many resonances with the success of IT based enterprises in the 1980s and 1990s, but these resonances also extend to "traditional" product-based businesses. In a recently published book, CK Prahalad notes interesting developments in the personal care product industry as a result of the convergence of markets and technologies.

> "Revlon, L'Oreal, Unilever, Shiseido and Procter & Gamble are moving up the value chain in their businesses. With an ageing population, skin creams are increasingly incorporating an anti-ageing ingredient (such as Retinol) and shampoos may incorporate a hair growth hormone (such as Rogaine). Personal care products are becoming more sophisticated and are beginning to utilise the knowledge bases that have long been familiar to the pharmaceutical industry." (*Management 21C: New Visions for the New Millennium*, October 1999)

Again, convergent technologies play a key role, as well as the readiness of the environment to buy. Without those factors, and without instincts for and awareness of those factors, the application of knowledge has no lasting value whatever. More interestingly, the successful enterprise often has to recognise and acquire knowledge that *it does not currently have*, precisely *because* of the context it finds itself in.

The so-called knowledge-rich enterprises, the computer pioneers of the 1980s, lacked all of the supposed criteria for economic success: no capital, no money, no assets - only their brains and a good idea. Actually, they didn't know much then either. If they had known how little they knew, perhaps they wouldn't have persevered.

There is an element of cognitive dissonance involved in the venture capital investments which were all the vogue in hi-tech startups of the late 1990s. These are the high roll gamblers of the old west, transformed into suits and cigars financing the workshops of Silicon Valley. The market itself had become risk pursuant - the stock value of hi-tech "knowledge-rich" companies far exceeded their asset value and even their balance-sheet performances. The phenomenon is not new.

> In 1878, the astonishing and impertinent Thomas Alva Edison procured $50,000 in venture capital investment for his Electric Lighting Company from the leading financiers of his day, W.H.Vanderbilt, J. Pierpoint Morgan and Western Union.

> The electric lightbulb was still just an idea in his mind, he had no idea how to make it work, there was no working prototype, and the leading physicists of the day said that his idea was impossible.

What lies behind this audacity? Not knowledge as we traditionally perceive it, although *attitudes* to knowledge figure in the equation. There is a key difference from gambling in the minds of these financiers, the robber barons of old, and the robber barons of Silicon Valley: in gambling, the odds are stacked in favour of the house. In venture capital risk, the odds are not especially stacked against you, and the financiers who have already made their millions see a percentage in the wild card - especially if the wild card suggests revolutionary (and profitable) changes.

We, the respectable people, tend to ignore this. We believe in rationality: we believe it all has to be worked out in advance, and the numbers have got to work. The risk takers and the visionaries work completely differently. The point is this: it is not what they know, that makes them successful, it is *desire to know*. This is the fuel that tells them what they need to know, and enables them to create new knowledge.

The curiosity factor

Creating the knowledge inventory or knowledge reservoir is therefore not enough: to be truly knowledge based, you need to incorporate knowledge-creation skills into your organisation: and that means building space for entrepreneurs, visionaries, mavericks, risk-takers and explorers within jobs, teams and projects. It also means building in buffers to protect this space from the more institutionally and rationally minded of us, and the skills effectively to mediate the conflicts and frictions that inevitably occur.

This is a much, much harder task than building groupware, a data warehouse or a just-in-time knowledge delivery system. And it is a task that we persuade ourselves that we have completed when we send our staff on creativity or communication skills courses. We haven't. We haven't even begun to scratch the surface, and the organisations that have the greatest chance of succeeding in this are always the startups, and always the entrepreneurs.

Andy Grove recognised this when he spoke of the "Cassandras" in an organisation who call out warnings about the "strategic dissonance" between what the organisation knows, and what is happening in the environment around that organisation. Sometimes, he says, you have to abandon your knowledge if the signals you are getting tell you that it no longer works. (*Only the Paranoid Survive,* 1996)

Although we try to persuade ourselves that the business world is in daylight, in reality we are driving in the dark, seeing no further than our headlights. To navigate we read the signs and follow the markings as they appear briefly before us. To navigate well, we also need imagination, maps, and a radio tuned to the right station. We may not be able to see what lies beyond the edge of our lights, but it is important that we are *curious* about it.

13
Buying into Risk

Investors love risk, managers dislike it. Risk provides an alternative to making money by the sweat of your brow. Putting money into a risky situation is what makes profit. For those who doubt this, take a look at the insurance industry – a business wholly devoted to the assessment and conversion of risk to financial premium.

> In Singapore alone, about 35,000 people (almost 2% of the workforce) are either wholly or partly employed in this industry. The assets of the industry in Singapore grew by more than 50% in the two years of the Asian financial crisis to S$31 billion. Over the last five years of the century the annual growth rate averaged 16%, reaching revenues of almost S$8 billion in 1998. There is a lot of money in risk.

The whole story of business is the story of investors putting their money into unknown, uncertain, unexplored territories, in order to multiply it severalfold. Investors need managers, because managers are trained to minimise risk. Having created the risky context for their money, their concern is now to make sure it is as well-protected as possible in its search for added value. Managers put systems into place, streamline processes, create risk-avoidance procedures, and steer the investor's money into safe harbour.

Managers need risk much as doctors need sickness – without it, they had better find something else to do. They don't like it, but it defines everything they do. Accountancy is about minimising the risk of loss of assets; decision-making skills are about minimising the risk of error; meetings and teams are partly about spreading responsibility for risk of failure; management information systems are about minimising the risk of unexpected and dangerous environmental factors.

Risk and knowledge

Business is a kind of guerrilla warfare between the creation of risky environments by entrepreneurs, innovators, investors and competitors, and the minimisation or taming of that risk by managers. And it needs to be that way: risk pursuance is pure folly unless it is accompanied by risk avoidance.

Let's go back to the insurance industry. The key tool of any industry player in the business of risk is knowledge. No player will agree to translate risk into money unless they have an actuarial analysis into which this policy fits. Knowledge is the single most important factor in the management of risk – whether it was the possession of accurate maps and local guides for the Portuguese explorers of the 15th and 16th centuries, experience and judgement in military and political affairs for the colonial powers in the three hundred years of the East Indies trade, or research into time and motion studies for massive productivity growth in mass production manufacturing at the beginning of this century.

Loss from		**Example**	**Knowledge Model**
FAILURE	Predicted value adding situation fails to materialise. Managers get their predictions wrong, investments suffer.	Use of competitive intelligence, market research, economic and demographic data, research and development.	Explicit to Tacit, Tacit to Tacit, experience and judgement informed by data, information and knowledge to make predictions.
FOLLY	Predicted value adding situation does materialise, but execution of company's positioning is flawed, through poor systems, communication or employee preparation and competency. Company is not positioned to capture value from new situation, even though it had been anticipated.	Skills and competency development in project management, leadership, communication, ITC infrastructure investment, systems and workflow design.	Tacit to Tacit, involving judgements about adequacy of competencies and expertise, and transfer where necessary. Tacit to Explicit in systems design, publication and training. Explicit to Explicit in learning from best practices.
FRAUD	Movement of money and assets is subject to criminal or malicious diversion.	Robust accounting and audit procedures, delegation of financial responsibilities, separation of ordering from payment responsibilities.	Explicit to Explicit in implementation of public, transparent systems. Transfer of best practice models.

The manager minimises risk and enhances safety in three key areas. She puts into place systems that protect against losses from failure, folly and fraud. All three systems use knowledge to provide against the negative impact of risk.

Even a cursory examination of the knowledge models designed to minimise risk suggest immediately the differences between the mature organisation (which has most if not all of these models and structures) and the entrepreneur (who probably has few of them). It brings home immediately a key fact. *The avoidance of risk itself invokes heavy costs*. The mature organisation spends a great deal of money on infrastructure, checking, doubling up on procedures, documenting, security, and decision-making. It is, consequently, slow and unwieldy. Every decision that moves money has to go through several heads and processes, contrasted with the entrepreneurial small enterprise which is nimble, moves resources quickly and effectively, but has a higher risk exposure, and will fail more often in what it does.

This presents us with an apparent paradox: *both the avoidance and the pursuance of risk* impose costs – either through the systems and time invested in risk avoidance, or in the higher percentage of failures associated with riskier activity.

Risk and branding

Why does the large mature enterprise spend to *avoid* risk, while the small entrepreneur spends to *pursue* risk? The answer probably lies in the association of success with brand strength. The possession of a strong brand reputation is more likely to warrant investment in risk avoidance. In relation to the value of the brand and the potential losses to that brand value should it be associated with repeated failure, such investment is probably justified. This issue is especially important in the case of publicly funded and government organisations, where the trust of the population in the integrity of the structures for transferring value is highly sensitive to loss and failure. Risk aversion is strongest in civil servants for good reasons – when brand value of publicly funded bodies is compromised by risky behaviour and loss, the implications become political as well as economic.

The small entrepreneur, on the other hand, usually has limited brand value to protect. Her loss is limited to the investment and to her personal reputation.

> Successful entrepreneurial businesses that undergo rapid growth generally begin to establish brand sensitivity and risk aversion in parallel – the poor performance of Dell computers in 1993, for example, resulted in Michael Dell hiring experienced veteran managers to build systems that were capable of delivering at a greater level of security and scale.

On the other side of the coin, in the new environment, large companies are beginning to perceive that they are too reliant on existing systems for minimising risk, and they are dangerously inexperienced in engaging with risk and devising new ways of dealing with it. In a business landscape that poses new and unforeseen risks this could be a fatal weakness.

Procter & Gamble's (then) new CEO Dirk Jager told *Fortune* magazine in early 1999, "Everybody is always worried about taking risks because nobody likes to fail. But you have to celebrate failure." He didn't mean that, of course: he meant, as most "failure is good" rhetoric does, that you can learn from and celebrate intelligent failures, just so long as you succeed in the end. Procter & Gamble then went out and spent a year demonstrating to the business world just how inexperienced it was in playing with risk – its latest embarrassment before Jager was ousted – a failed hostile takeover of Warner-Lambert and American Home.

Pursuing risk is too easily associated with beating one's chest in public – it is good to remember the bespectacled actuary behind the vast profits of the insurance industry. Knowledge and experience do count.

14
The New Landscape of Risk

Why then are we suddenly asking our managers to pursue risk? It is not that the manager's job in the new economy has changed, and risk of itself suddenly brings automatic rewards. It doesn't. We are as prone today to loss through folly, fraud and failed predictions as we ever have been – indeed, in these volatile times, some would say more prone. We desperately need sound and prudent management of the risks we are pursuing – and that means risk aversion on a scale to balance the risk pursuance of the markets.

The problem for managers is that the landscape of risk has changed. By the year 2000, four years of bull run on the equity markets had reduced sensitivities to the downside of risk, and increased appetites for the amount of risk that investors would tolerate. Higher risk means higher profits from the investments.

In the 1950s the investment debates were about levels of risk tolerance between the relatively safe bond markets and the perceived riskiness of the equity markets. History has proven the winners correct: the equity markets have consistently outperformed the bond markets in the UK by a factor of 4.5% over the past 100 years.

Higher risk equals greater profits, so history seems to say. Now those very same blue chip equities were being challenged by the upstart white hot high tech and internet stocks, and what was adventurous fifty years ago now appeared unexciting.

History, from the Romans to the Mongol horde, to the Dutch East India Company, also teaches us that changes in the technology of communications infrastructure brings about rapid economic shifts, creating enormous opportunities for profit.

> Advances in maritime navigation techniques and technology opened up the Asian and American markets for European exploitation in the 16th to the 18th century. The railways, followed swiftly by the telegraph, prepared the world for the economic and cultural domination of North America by creating around its rich natural resources a vast production and transportation network, that in turn cultivated an enormous domestic market, that in turn nourished a mighty industrial machine hungry for new global markets.

The new information infrastructure

As Alan Greenspan pointed out at the peak of the new economy boom, the mid-1990s saw a new explosion in connectivity. This time it was an information infrastructure, as the technologies of computing, telecommunications and standardised networking protocols converged into a critical mass of global networks capable of real-time transmission of data from any point to any point. Real-time global access to information completely changes the traditional business models for risk.

Warehouses

Warehouses are expensive items to buy, build or rent. Their management involves all sorts of processes for getting things in and out of them, as well as accurate and timely inventories of their contents. Warehouses, as with any storage systems, are an insurance policy against the risk of not having what you want when you need it.

> Michael Dell proved that you can use networked information technology to keep your inventory on the road, cutting out an entire stage in the business cycle: storage and retrieval. The consequence of this is threefold: reduced costs, greater speed, higher productivity.

Gary Beggs, vice president of Yellow Freight in the United States told the *Financial Times*, "We've become moving warehouses." Yellow uses a logistics software called Sysnet, designed for it by Princeton University. Shortages of trucks and containers in one region and oversupplies in another can be addressed at the click of a mouse.

Airlines use similar software to keep their US$50 million aircraft in the skies, where they are earning money, and not sitting on airport aprons, depreciating by the minute.

Tendering and procurement

This is a complicated and expensive process for any large organisation. Much of the complexity and cost is associated with reducing the risk of poor match of delivery against order specifications, or the risk of getting a worse price than your competitors thus eroding your margins. Trust, authenticity, and assurance all cost money. In the 1990s automated supply chain management became the watchword for manufacturers and retailers, migrating in early 2000 to a new application for e-commerce – business to business electronic procurement marketplaces.

> Car manufacturers Ford, DaimlerChrysler, Renault and Nissan Motor have all set up shared electronic marketplaces, as have retail chains Sears and Carrefour, followed in rapid succession by 11 of their global rivals.

Management information systems

These are designed, in large organisations especially, to overcome the risk of sub-optimal co-ordination of resources, assets or people, and to avoid the risk of poor decisions in response to environmental change. We pay managers, executives and admin staff a great deal of money to gather data, create, check and distil documents, and transfer them to designated points at designated times. Networked organisations with real-time access to customised views of the state of play in any areas of the organisation's activity are not far away – some of them are among us already.

The greatest risk

Networks, connectivity, bandwidth and standardisation of information transmission protocols have evaporated many of the traditional risks around which our management systems have been created. The systems we created were borne out of a lack of immediate knowledge, and an inability to respond quickly to change. IT has given us much greater speed of response, and much higher levels of immediate knowledge.

And yet, although the traditional risks have gone, we are patently not without risk. The truth is, we are in a landscape we do not fully understand, and our ignorance itself poses the greatest risk. We are, like the Portuguese explorers hunting down the fabled Spice Islands, navigating new domains.

> When Julian Robertson bowed out of the Tiger Fund, he stated our common situation very clearly: "There is no point in subjecting our investors to risk in a market which I frankly do not understand." This was an uncanny echo of Warren Buffet's admission around the same time that investing in the new dot.com marketplace was "outside our circle of competence."

There are clearly fortunes to be won and lost in this new marketplace, as there were with the advent of railways in the 19th century. We see them all around in the new millionaires of Silicon Valley.

> After four years of the strongest bull run in recent history, by the year 2000 millionaire households made up 4% of the US demographic picture.

As in the chaotic rise of the railroads in the mid nineteenth century, we see today speculative fevers, fraud, insider trading, stock options, stockprice manipulation, bankruptcies, small branchline companies being bought out by larger, consolidating and integrating concerns, irrational shifts in the financial markets. We also see new money, in the shape of the Vanderbilts, Goulds, and Morgans, overcoming the old. We also see the enormous volatility in the markets accompanied, unexpectedly, by rapid growth in productivity and output.

From 1860 to 1900 the annual rate of growth averaged 4.3%, and in the midst of financial chaos was the longest sustained period of economic growth in America's history.

> US Federal Reserve Board chairman Alan Greenspan remarked on a strikingly similar picture in the United States economy of today: amidst "discontinuous shifts in economic structure" the economy was achieving in the last five years of the 20th century an average of 3.5% growth in output per hour, almost double the average for the previous 25 years.

Analysts point to a "step change" in the economy – enabled by the technologies of information and connectivity. Risk too has changed, and until ignorance of this new landscape can be contained by experience, judgement and knowledge, we will be uncertain as to our ultimate destination: whether it be in the camp of the high rolling winners, or the bankrupt losers, or the middle ground in between. We return, as investors and workers in the late 19th century did, to the gaming tables of chance and speculation, risking far more on the rolls of the dice than we ought to, but knowing that if we do not roll, we cannot stay in the game.

It is dangerous to stay away from the risks of the new environment and pretend they don't exist, because the risk of being excluded from the new economy means even greater loss. The risk averse manager understands this, and goes for the lesser risk, engaging with the new landscape, charting it and navigating it for her investors, building the body of knowledge and new systems that will bring what is now a game of chance to the status of a profitable risk management enterprise.

Engaging risk

The ability to set present systems aside, and work from first principles is paramount for success in this new landscape. Risk from failed predictions still needs competitive intelligence – the new tools need to be learned, and new systems devised for disseminating this knowledge to where it needs to go. Risk from folly and poor implementation still needs a focus on developing competencies and skills in oneself and others – it is just that they are new skills and they need to be distributed differently. Risk from fraud is as real today as it ever was, but it comes from new directions as well as old, and the emerging technologies of security, authorisation and financial control need close technical supervision.

Should managers pursue risk? No, they should seek to minimise it as they have always done. That's their job. But in order to do that successfully, they need to engage with it, understand it, tolerate failure only to the extent that it teaches them about success, and ultimately seek to master it.

Should companies seek to pursue risk? All companies engage with risk – business is after all, the making of money through engaging with a risky environment. Companies and organisations can only engage with the level of risk that they can afford, however. Big, mature organisations need to learn how to engage with risk and assess it wisely. Smaller, nimbler, entrepreneurial companies need to learn skills of risk aversion if they are to accommodate growth and brand building.

A matter of survival

The question facing both organisations and managers today is a question of knowledge. Knowing the risks means knowing the opportunities. Knowing the opportunities coupled with the capacity to act swiftly means the creation of value. In a connected world, risk has changed its face – our learning curve needs to be steep.

On August 5, 1858, the first transatlantic telegraph cable between Europe and North America was brought ashore in Newfoundland. It failed within a month, and it would be seven years before a successful working cable was in operation, but this date marks our last major shift in the boundaries of business risk. The spectrum of risk accompanying commercial and political communications across the Atlantic was suddenly brought low by the possibility of instant messages and instant news, transmitted independently of the risks and delays of transatlantic travel. It was now possible to do real-time business with Europe, without any of the risks of hazardous transatlantic travel, and knowing that a shipment would arrive only six weeks later than the known market when ordering, instead of three or four months. One of the first messages transmitted was an instruction to General Trollope countermanding previous orders to send two regiments of troops back across the Atlantic to be redeployed in India. The Indian Mutiny had ended while his original orders were on their way to him by sea, and the telegraph message saved the British government £50,000. Few in the celebrations realised that reduction of some risks meant increases in others. Keeping the lid on industrial espionage, leaking secrets from the 19th century's industrial powerhouse – Britain— to its 20th century successor, the United States, was suddenly much, much harder than sending Customs agents to search the luggage of suspicious-looking American travellers as they embarked for New York.

Two hundred and thirty eight years earlier, in December 1620, the Pilgrims had stepped ashore at Plymouth Rock, 500 miles to the south of where the transatlantic cable was winched ashore. Their goal was to start a new life by colonising and taming the New World, but they were woefully equipped for such a risky endeavour. Their profile was that of small town petty bourgeoisie, not especially skilled in agriculture, hunting, fishing, botany or construction. They numbered two tailors, a printer, several merchants, a silkworker, a shopkeeper and a hatter amongst them. Their cargo of shoes and clothes, sundials and snuffers, trumpets and books suggested they were moving house rather than creating a new settlement in virgin, undeveloped territory, with no European neighbours for hundreds of miles in either direction. They had neither the tools nor the skills to manage the risks and challenges of their new environment. In their first six months, 50% of them died.

To the company executive of 2005, it might seem that today's companies, with their order-forms in triplicate, documents travelling around organisations accumulating multiple approvals and signatures, handwritten stock control forms, printed bank statements several days old arriving in the mail, training budgets getting larger the further up an organisation you go, strategic management decisions being made by a Board of Directors based on last year's financial performance, are all peculiar features of an obsolete worldview, and especially inappropriate to the new risks and dangers of a wild and virgin territory.

In the *new* New World, the new executive will wonder how so many of the managers of 1999 survived, when their perceptions, skills and attitudes were so evidently unsuited to the new environment.

As with the Pilgrims, so with us. It is a matter of survival. We had better learn fast.

15
Fast Technology, Slow People

I have wanted for some time to write about friction. We are exhorted to operate in internet time, and great efforts are expended in speeding things up "to make them more efficient". Yet, at the same time, almost unnoticed, we are also trying to slow some things down. This disparity between speed and slowness is expressed most cogently, to my mind, through exploring the notion of friction, that point of contact that turns speed into slowness and produces something else in the process.

Some friction is good, some friction is bad. In the next two chapters, I am going to consider two types of "bad" friction: human stress, and the constraints of the physical world. I want to do this in order to explain a few pressing questions: Why is the Nasdaq still weak? Where has all the venture capital gone? Why is cash burn now a dirty word and first mover advantage suddenly unconvincing? Why, after ten years of internet technology, downsizing, mergers and acquisitions, ERP systems, business process reengineering and constant innovation, are big companies still struggling to survive?

The dissociation

In 1841, the Great Western Railway agreed to extend an experimental telegraph line from London Paddington to Slough, a distance of 20 miles. Although exhibited (for a fee of one shilling) as a marvel of modern technology it was little more than an odd showpiece to the general public who did not much understand what it was for. For the next three years it acted as little more than a device for the railway to send advance notice of the arrival of its trains.

One day in August 1844 that started to change.

> Throughout 1844, two London pickpockets, Fiddler Dick and Oliver Martin had a nice little scam going. They would loiter around near the railway terminus in London, and pick pockets. When discovered, they would jump on a train, and travelling at 40 miles per hour, evade capture through using the fastest form of transportation available at that time. On 24th August, however, a bright station

> attendant decided to use the largely decorative telegraph system to wire ahead the descriptions and carriage location of the two thieves. At Slough, Fiddler Dick and Oliver Martin were arrested by waiting police.

Another more notorious case happened early in the following year.

> John Tawell, a respectable Quaker from London, travelled to Slough on New Year's Day 1845 to poison his mistress with prussic acid. Instead of expiring gracefully at his feet as Victorian mistresses were supposed to do, Sarah Hart let out a bloodcurdling scream and collapsed in noisy agony, thus alerting the neighbours. Tawell fled the scene, and was observed boarding the 7.40pm train for London. The telegraph was used to wire ahead Tawell's details and location, and he was picked up at the other end, convicted and hanged for the murder.

These two incidents, at two ends of a 20 mile telegraph line, mark a critical stage in human history: the point at which complex information began to move faster than human beings could travel. Up till then, information and knowledge travelled with their carriers, and a society's speed of reaction, anticipation and response depended on systems of transportation operating in equivalence with systems of communication. From this point on, technology would fuel a gradual dissociation between the speed of information and the speed of human beings. The stresses and strains of adapting to change are directly related to this dissociation, and the struggles of human beings to keep up with the speed created by technology.

Already in 1835, Bavaria's College of Medicine was warning would-be train passengers that the unprecedented speeds would result in brain damage (*delirium furiosum*). By the 1860s there was a voluminous medical literature on the effects of train travel, much of it to do with the negative psychological effects – especially anxiety – created by travelling at speed. In 1868, Alfred Haviland published a monograph entitled–*Hurried to Death*. We treat these tracts somewhat dismissively now, forgetting the impact of shrieking, grinding, smoky and uncushioned transport as compared with the leisurely rocking of organic transport systems. Yet 100 years later, we find Alvin Toffler defining "Future Shock" as "a human response to overstimulation", resulting, variously, in apathy, anxiety, hostility, senseless violence, illness and depression (*Future Shock*, 1970).

Why should the dissociation between speed of humans and speed of information be a problem? If we were simply passive observers of information flows it would not be. But information has to pass through human heads in order to make value. Human beings are insatiable information consumers and producers. And we compete on it. We pick up an idea here and turn it into something else, and make money from it. Someone else sees it and follows suit, wiring the idea her own particular way, and makes more money. Suddenly my market is under threat and I need new information, new knowledge, new ideas. The more swiftly information can propagate, the more access to new knowledge, information and ideas I can get, and yet, conversely, the more competitive and uncertain my environment becomes.

When money stops being a precious metal and turns into a particular form of information transmitted down wires, and when regulatory controls are relaxed, my problems multiply. Money moves fast, and I must move faster to keep up with it. When AOL sets up its global technical support centre in Subic Bay Philippines, jobs disappear from San Jose California. I had better not be in San Jose if I am an AOL technical support call centre operator.

Human friction

Why is change management so hot in today's climate, with an estimated worth of over a billion dollars for the software industry alone? Friction. In the frictionless environment of cyberspace, information moves at lightning speed. When it enters human heads, it must slow down to human speed. What happens when a fast moving object meets another object and slows down? Kinetic energy gets turned into something else. In the case of electrons travelling along copper wire, or a meteorite entering the atmosphere, you get heat.

In the case of information travelling through human synapses, you get stress. Human beings like to digest information before they decide to act. The faster the information comes, the more stressful the experience of deciding and acting is. And humans being what they are, when they get stressed they do all sorts of inconvenient things like slowing down, becoming less cooperative, seeking help and advice, seeking social support, becoming less motivated, taking longer to come to decisions, devising irrational explanations for the barrage of pressures coming at them, pretending it isn't happening.

If this seems like a fanciful explanation, look at the facts in research from A.T. Kearney, Gartner Group and others.

> Fewer than 50% of companies undergoing restructuring, delayering or downsizing realise lower costs or get higher productivity as a result (1992). About 80% of TQM initiatives fail to achieve tangible results (1997). Roughly 90% of Business Process Reengineering projects fail to meet targets set (1997). Only 25% of companies implementing ERP systems are successful within 18 months, and after three years the total is still only 30%. A fifth of all of ERP programmes are aborted (1997). Only a third of all knowledge management programmes achieve success (1997). Over half of all Sales Force Automation projects fail outright because they overrun or are never completed (1999). European managers consider that only 20% of their change programmes could be considered successful, 63% had made a temporary improvement that they failed to sustain, and 17% had failed completely (1999). Almost 60% of all mergers and acquisitions fail to achieve desired goals, and mergers based on short-term goals actually destroy value (2000).

The "problem" every time? Rarely the technology, usually the people. Maybe it's worth taking friction seriously.

16
Slow People, Slower World

In the previous chapter, *Fast Technology, Slow People*, I looked at the growing separation between the speed of digital information and human speed. When information and humans meet at vastly different speeds, you get friction, manifested as stress, and slow response times in an environment of lightspeed urgency.

There is another dimension to the discontinuity between the frictionless environment of cyberspace, and the friction-laden environment of the physical world, and it is closely related to the world of investment, speculation and risk. To see it in action, we need to travel back in time to the Paris of 1720.

> It is December 1719, and we are in the rue Quincampoix, a dirty, narrow, crowded street which forms the heart of Paris's financial district. Shares in John Law's Mississippi Company, originally issued in May 1719 at 500 livres, are now selling at 10,000 livres, a twentyfold increase, and this is the only place you can get them. Traffic became so congested earlier in the year that barricades have been put up, and only pedestrians are allowed now. Speculators have taken over houses and charge extortionate fees for investors to gather, watch the trading, and gather information and gossip. An estimated 250,000 foreigners are in Paris to play the markets. The French government owns fully one third of the shares in the Mississippi Company. At the beginning of 1719, France had a national debt of 2 billion livres. At the end of the year it has a surplus of roughly 3 billion.

We are, of course, in the middle of the speculation frenzy of the Mississippi Bubble, with 7 billion livres in outstanding shares against only 2 billion in ready money. At a leverage of more than 3 to 1, it was bound to collapse, and it did, dramatically, resulting in the ruin and flight of John Law. At the beginning of 1720, investors made wealthy on the scheme began to hear the truth about the dismal state of the colonies in the New World, and they started to cash in their winnings, turning the shares into property, silver and gold. Many of them, touched by speculation fever, took their money to London, where the South Sea Bubble had another nine months to run its course to ruin.

Unlike the South Sea Bubble, the Mississippi Company had a real value proposition. The Company was formed to exploit the vast resources of the Louisiana Territory, over 800,000 square miles of territory from New Orleans to Canada, rich in forests, land and minerals. Since 1717, glowing stories of the potential for wealth from this new world had circulated in Paris. There was gold, silver, emeralds, and fertile land. All that was lacking was the human talent to exploit it. All that was true, and with rich oil reserves to be discovered two hundred years later, the territory would be even more valuable. As a *really* long term investment proposition, you couldn't lose.

Friction and time

The problem for John Law was that the real world could not be manipulated quickly enough to create tangible value and generate a productivity that could satisfy his unprecedented leverage of 3.5 livres outstanding for every 1 livre in the bank.

> In the midst of glowing propaganda throughout 1719 and the early months of 1720, the reality was that New Orleans, described in Paris as a gracious city of 800 homes, was actually a desolate village comprising four houses, and the many thousands of settlers were dying in droves from dysentery, hunger, yellow fever, scurvy, and Indian attack. The Mississippi Company tried to remedy the situation, fitting out ships, and forcibly transporting thousands more of unwilling settlers throughout 1719 and 1720. Between half and three quarters of them died in the process.

What was happening here? Friction.

This time, it is the dislocation between the actual capacity of the physical world to generate value and people's beliefs and desires about that capacity. What fuelled the investment frenzy throughout the summer and autumn of 1719 was lack of knowledge. Ignorance raises risk, higher risk means higher rewards, but lowers your chances of getting them. In the case of the Mississippi Company, however, the ignorance was not about the capacity of the territory to create value. The ignorance was about *how long it would take*.

> The 14 American States that eventually emerged out of this territory did not see their greatest population growth until the 1850s, had their greatest manufacturing investment growth and farm land value increase in the 1860s, but did not reach their productivity peak until 1930, when with only 23% of the USA's landmass

and 19% of the population, the Mississippi Company's territory was producing almost 40% of the USA's output.

If John Law's investors had been prepared to wait 210 years, they would have had a really good deal. If we translate John Law's story to our present situation, we can see some strong parallels. We, too, are investing in a new continent. We know very little about how it really works, or about the true value of what it could be worth. All we know is that it is BIG, and we fear that if we don't get in there at the start, we will lose out. What we are beginning to realise, is that we need to have really big lungs and really deep pockets. How long can you hold your breath?

When human desires hit the slower moving, six thousand million trillion metric tonnes and complex ecology of planet earth, friction must ensue. In human terms, this is expressed as risk, and it requires both stamina and courage. Traditionally, risk is calculated in two dimensions: the first dimension is that of probability, and the second dimension is impact or reward. As probability goes down, usually impact and reward go up. What John Law's case suggests is that mapping against a third dimension of time could have interesting results. If you want to assess the worthiness of an investment, and as time increases, both probability and impact go up, you are on to a good thing – so long as you can afford to hold your breath. For the sake of this argument, I am assuming that you are not approaching the stake as a gambler for short term gain, because entirely different rules operate here.

Levels of friction

It does not seem too foolish to conclude that we poor humans have two major dissociations to struggle with: the dissociation between the speed of information and the natural speed of humans in dealing with information; and the dissociation between human will and the ability of the physical world to react to that will. Hence, there are two levels of friction that businesses and investors must consider carefully.

(1) What is the friction quotient in your people's ability to deal with technological change and knowledge flows, and hence what levels of resistance and stress can you anticipate?

(2) What is the friction quotient between the ability of the tangible world to reward your investment, and your precise intent, including your time-frame?

Why then are we struggling in this new economy? Because technologically enabled companies are faster than human beings, but must struggle through them; because people are faster than the physical world, but must manipulate the physical world to achieve their ends. Change management, communication skills building and the focus on leadership and empowered teams are all attempts to speed up people and reduce stress. The retreat into selling knowledge and services rather than tangibles is an attempt to escape the ponderous intractability of our home planet.

One of my intentions in this chapter was to examine the notion of "bad" friction. However, the reality is that the mismatches between the speeds of information, human beings and the world only become a negative if the baseline is taken to be the speed of information. If we take the speed of humans and of the world we live in as our pace-setters, then the fluidity and speed of information only has an upside. In future chapters, I shall be seeking to explore the upside of friction – why slowing things down creates value and opportunity, and spreads risk.

17
Lower Friction, Higher Risk

> In 1996, Professor Mohanbir Sawhney of Kellogg University wrote an acclaimed paper called "The Death of Friction". In it, he pointed out that in the USA, the average new car costs $20,000. Of this, fully one third is spent on promotion, distribution and dealership incentives. At 15 million cars being sold per year, removing the costs of the distribution networks would save over $100 billion in friction.
>
> Sawhney suggests that the rise of internet-enabled "metamediaries" who assemble integrated packages of information and services from a crowded, confused and fragmented marketplace could cut out many of the information components of this distribution network, creating at least $10 billion of value. "Metamediaries represent the death of friction," he says, "and the birth of frictionless commerce."

Sawhney is not alone in the celebration of lubricated information networks: Christopher Meyer and Stan Davis of Ernst and Young produced a book in 1998 called *Blur: The Speed of Change in the Connected Economy*, closely followed by Bill Gates in *Business @ the Speed of Thought* in 1999.

And the enterprise software vendors are racing for the killer applications that will give them access to this Holy Grail of frictionless commerce.

> Database vendors such as SAP and Oracle have moved into enterprise resource planning applications, and now they are extending out of the back office and into the B2B marketplace with software or alliances for internet exchanges and netmarkets that talk directly to back office inventory, procurement and sales.
>
> "Our vision is for the procurement guy to be able to see three steps up the supply chain into the inventory of his trusted trading partners," says a consultant from an exchange software vendor. "The potential savings are tremendous. No more thirty days inventory as a hedge against late deliveries; now you are talking about truly just in time supply chain management."

It is not too fanciful to use the metaphor of the Holy Grail. Enterprise Resource Planning systems are still a vision of connectedness and integration more often than they are a reality, despite the billions spent on them.

The old economy worker

The vision is, of course, the removal of friction: the warehouse space, employees and management systems at every step along the supply chain pruned down into a constant flow of goods and materials on the road, arriving just in time to be used. Removing friction releases cash for the enterprise, and information flowing along frictionless wires holds the key to removing friction from the supply chain.

Hold that image. Interrogate it, along with Sawhney's vision for releasing $10 billion of value from the US car sales process. Aren't there people in there too? Warehouse workers, clerks, car dealers all get written out of the script in this new frictionless scenario. Isn't this also then a question of releasing billions of dollars of value from current workers in the old value chains and transferring those dollars to IT managers, senior executives and shareholders? Gaining leverage on resources to create value is not just a question of creating value: it is also a question of transferring it. And we can see exactly that happening at ground level.

> The Singapore Department of Statistics reported in December 2000 that the unemployment rate for unqualified workers had soared from 2% in 1990 to 9% a decade later.

Removing friction from the value chain removes people as well.

I suggested in chapter 15 that people generally are friction contributors in the globally networked business, but now we see more clearly a separating out of old economy worker from new economy worker. We also get a general sense that whole swathes of current expertise, experience and contribution can be made irrelevant by clever software and integrated networks, and they *will* be made irrelevant by senior executives who are hard pressed by the technology arms race and the absolute imperative to stay competitive. Discarding the old to create a larger pool of value for the new breed, suggests that the value chain exists only for people with certain qualities – an idea we shall revisit later.

The removal of friction from economic channels and spaces has provided competitive advantage to innovators throughout human history, and until recently this has revolved around innovations in the technologies of transport. The first societies to use the wheel were able to sell more of their produce further afield (economic influence), and they were able to supply their armies at greater distances from home because they could carry more food and weapons per unit of horsepower (political influence).

> The Romans built their empire on the back of their straight, flat roads, and commerce followed hard behind the legions.
>
> The 15th century Ming Court wielded economic and political power as far afield as Sri Lanka and Africa through innovations in its naval technology, and as its influence declined, the Portuguese opened up new eurocentric markets from Asia to the Americas through innovations in maritime navigation.

The competitive edge

Interesting things happen when you remove friction very quickly. At the beginning of the 13th century, the Mongols, a nation of less than one million people began an expansionary drive across Asia and into Europe that brought them, within 50 years, domination of the largest land empire the world has ever seen. They out-competed their target nations through ruthlessness and psychological warfare, but above all through their skilful use of the horse to provide the advantage of speed. Their competitors had horses too, of course, but the Mongols added a process to the reduced friction of riding on horseback. Their real competitive advantage? Their diet.

A nomadic people, they lived on the products of their horses – mare's milk, blood and meat. Their horses lived on the grass of the steppes that formed the avenues for their invasions. Unhindered by supply chains, therefore, their armies moved faster than news could travel, covering up to 50 miles a day. In extreme emergency, their messengers, using a post-horse system, could travel 1,000 miles in four days.

And while history might suggest other reasons for the ultimate frustration of the Mongol hordes, it is an interesting coincidence that their remorseless advance was stemmed and finally stopped when they tried to use processes and channels beyond their sphere of special competence.

> At the battle of Ain Jalut in Palestine in 1260, the Mongol army was defeated for the first time at the hands of the Egyptian Mameluks – in desert country, far from the greenery of the steppes. An attempted Mongol invasion of Japan by sea in 1281 failed disastrously when a typhoon hit a fleet containing 140,000 men.

Yet behind the all-consuming wave of the Mongol armies, with superb pragmatism, the Khans established secure roads and administration systems. Merchants found that they suddenly had an enormous common marketspace with relatively safe routes and incredible variety of products and markets. By the third generation, under Kublai Khan, fabulous wealth had accrued to the rulers of an empire with such well-established mercantile routes and marketplaces. The Mongols revived the old Silk Route and policed it, making it secure and safe. While they pushed ever outwards at the fringes of their empire, within it, safety and stability prevailed. Europe became curious, a young man of Venice named Marco Polo accompanied his uncle and his father into a world of new business opportunities and different technology. It was not by accident that Venice led both the European Renaissance as well as the westward spread of economic prosperity into Europe until well into the fourteenth century. The concept of first mover advantage in a new marketspace was not invented by Amazon.com.

Friction and scalability

It looks like a classic economic success story: a new differentiator (diet plus ruthlessness) is deployed to give a group competitive advantage. They use it to conquer and command new space. They build routes and channels for commerce, and prosperity and new opportunity follow on a far larger scale than before. It is the vision of 20th century globalisation prefigured. But there is a more interesting story behind this one.

> In the early 1330s, there was a deadly outbreak of bubonic plague in China. It took 10 years to reach the Black Sea, the nexus of trade between Europe and the Mongol-established marketspace. From there it spread north and west simultaneously, like wildfire.

It was in the lower Volga River basin in 1345, following the mercantile routes north to the Caucasus and Crimea (1346), and west to Constantinople (1347). As summer ran towards winter in 1347, it raced through Alexandria, Cyprus, Sicily and Italy. The spring of 1348 found it entering the port of Marseilles, from whence it travelled north to Paris. By summer it was in London, and by autumn it was already spreading through Germany and the Low Countries. In May 1349 Norway fell, and the disease moved through the easterly trading routes into Eastern Europe and Russia throughout 1350 and 1351, finally meeting its Crimean cousin still heading north.

In the space of five years one third of Europe's population was killed, an estimated 25 million people.

Reducing friction creates more than economic opportunity. Introduction of new transport technology coupled with war and/or economic growth has always been followed by disease. Ease of movement spreads disease as well as value.

As European powers struggled for economic advantage in the new resurgence of wealth and technology in the 15th century, from 1495 to 1505 a syphilis epidemic raged out of Europe. Fostered by almost continuous war in Europe, it was first documented in Italy among Spanish mercenaries, spreading northwest through Switzerland and Germany, and reaching England and Holland in 1496. The Portuguese explorer Vasco da Gama carried it on his ships to India in 1497, from whence it spread eastwards into China and Japan, reaching both countries in 1505, faster than the Portuguese themselves could travel. They were not to reach Canton until 1520, long after their advance gift had subsided into a background disease.

Westwards expansion into the New World also carried a deadly passenger.

In 1520, smallpox arrived in Mexico with a Cuban slave in the invading Spanish forces. As they raced through the country searching for gold, the disease travelled faster than the Spaniards could slaughter the resisting population, so that chroniclers could not determine which was the worse enemy. An estimated 10 million Aztecs died.

And more recently, the following two cases:

In 1817, a cholera outbreak at the Kumbh Ganges festival in Calcutta travelled all over India with pilgrims returning to their homes, and started a 40 year series of epidemics that washed back and forth through Europe in infected water,

> following the same trade routes the bubonic plague had taken 500 years earlier. This time it travelled by steamship and train.
>
> A hundred years later, a new virulent strain of influenza broke out in an army base in North America. It travelled to Europe in the lungs of soldiers heading for the trenches at the tail end of the First World War. The death of friction in transportation had reached its final triumph. It took just a year to travel round the world, killing anything between 20 and 50 million people.

Perhaps, in our physically connected world, we will never again see ravages on such a scale. Yes, the more we travel, the easier it will be for disease will travel too, but there are very few pockets of humanity that have not had the chance to harbour and build resistance to a variety of diseases.

But surely the movement of friction-reducing efforts from physical space (transportation technology) into information space (connectedness and knowledge flows) cannot possibly carry the same risks the human race confronted as it built its physical economic networks. Think again. The whole point of reducing friction is to "release" the value embedded in the value chain for particular stakeholders. Essentially, the objective is to create more leverage out of the production process, so that smaller investment will scale to larger output. Friction inhibits scale.

What this means for today's globally connected business is that a mistake made in one scaled enterprise will be multiplied not across the local marketplace, but across the world. And it will be multiplied instantly. Some of that risk is still disease.

> Water supply systems established to promote convenience, hygiene and health in 19th century London actually transmitted the original cholera germs behind the 1854 epidemic. Blood transfusion services established to reduce the friction of access to adequate medical care played a large part in the transmission of HIV in the 1980s. The failure of the UK government to recognise the link between BSE and new variant Creutzfeldt-Jakob Disease in the 1980s meant that infected beef was entering the human food chain for over ten years until a worldwide ban was imposed on British beef in 1996. European complacency over their own control of the food chain resulted in simultaneous outbreaks of BSE in herds across Europe in late 2000.

In my next chapter, I shall explore how we multiply risk across our networks in perfect step with the success of our attacks on friction.

18
Who Pays for Risk in the Connected Economy?

In my previous chapter, I explored the notion that reducing friction by a quantum amount rather than incrementally, can produce devastating competitive advantage. The darker shadow behind the apparently positive tone of this tale was that when quantum leaps in channel lubrication were focussed on transportation technology, old risks of disease multiplied themselves in the enlarged market spaces into deadly epidemics. When, in the last decade of the twentieth century, we changed our focus to reducing friction through information technology, the nature of our risks also changed. It is not just disease that we facilitate through the lubricated, connected economy that we have built.

> In 1990, the French company Perrier recalled 160 million bottles from supermarket shelves after the product was found to be tainted with benzene.

In many ways that was a forerunner for a disturbing multiplication of such events. The more we scale our networks, the more we multiply mistakes, with devastating effect. And the more our competitors rely on economies of global scale, the more our margins are pressed, and the more we must seek new ways of lubrication. This is a strong incentive to tolerate small mistakes, and perhaps deny they exist.

> In August 2000, Mitsubishi Motors made a spectacular admission. For the past 20 years they had systematically concealed customer complaints about exploding airbags, brake failures and fuel leaks.

And in the brave new extended enterprise, it is also almost irresistible for contributors to make the mistake of pushing the responsibility onto another partner, recently demonstrated in the very public bickering between Ford and Firestone over who was responsible for the deaths of over a hundred people and injuries to hundreds more.

Let's examine a recent 18-month period spanning 1999 and 2000.

> In May 1999 Belgian food products were taken off Singaporean supermarket shelves when it was announced that some eggs, chicken and pork had been tainted with the carcinogenic dioxin at least since February. The following month

the safety of the food chain came under query again in a row between the USA and the European Community, which was resisting the import of beef and dairy products from cattle injected with bovine growth hormone. While increasing productivity, the hormone has been potentially linked with breast, prostate and colorectal cancer.

At about the same time, Coca Cola was forced to withdraw and destroy US$80 million worth of stock from four countries after traces of fungicide were found on bottles from a Belgian bottling plant. The company's arrogant response to initial consumer complaints of nausea and headaches did much to diminish its reputation.

In the midst of the 1999 Christmas shopping season, the European Union suddenly banned a wide range of soft PVC teething toys for babies, after concerns emerged about the safety of the chemical softeners used in their manufacture.

In May 2000 Toshiba paid US$1 billion in compensation to US customers for faulty floppy disk drives, sparking a row with Chinese consumers, to whom they had offered nothing.

Also in May, Scottish, Welsh and German farmers discovered that over a million pounds worth of oilseed rape crops would have to be destroyed because they had accidentally been sold genetically modified seeds that had not been approved for the food chain.

In June, the Japanese company Snow Brand gave food poisoning to over 15,000 people through a badly bungled health crisis, uncovering a whole range of quality, health and crisis management problems within the company.

In August, Firestone recalled 6.5 million tyres and provoked a US congressional enquiry into accident claims that went back to 1992.

In September 2000, Honda Motors recalled over half a million cars in Japan because of audio system defects and faulty engine oil seals, while Kraft Taco Bell recalled taco shells in the USA made with potentially allergenic genetically modified corn, and Mitsubishi Electric recalled 45,000 TVs in Japan after 66 recorded incidents of bursting into flames.

To the consumer the risk

Connectedness brings risk, because damage can hit everywhere in the chain at once. The jury is still out on the safety of mobile phones, but meanwhile there are 470 million mobile phone users worldwide, and that number is growing by about 50% per year.

The real question that concerns me here, however, is where the risk falls. And it is disturbing that in the connected economy the risk falls mainly on the end consumer. Yes, Coca Cola loses US$80 million of stock, Toshiba pays out US$1 billion, Firestone faces claims of more than US$2 billion on top of a US$500 million tyre recall charge.

But these are minor costs in relation to the revenue and marketspace advantages of the globally scaled enterprise. There appear to be two main laws in the risk accountability of companies that hurt their customers.

1. Law of the Few – relatively few cases are often amplified in alarmist ways.
2. Law of Forgetting.

> Perrier and Coca Cola still dominate the world markets, people still eat Belgian chocolates and eat beef without checking where it comes from, and Bridgestone came out of an even bigger tyre recall in 1978 to take 30% of the US tyre market in the year 2000 before the latest scandal hit.

It is not the body corporate that takes the greater risk when mistakes are multiplied across networks, it is the consumer.

While reducing friction in the economy removes the costs of moving between fragmented mini-economies and systems, it actually transfers those costs to the consumers in the form of risk. Along with your global banking, wireless access, 24x7 customer service, and global logistics and fulfilment, the new frictionless system also quietly distributes risk into your home. Looked at from this direction, the inefficiencies of the old, fragmented, bordered and territorial world actually operate as circuit breakers for risk.

> When Malaysian Prime Minister Dr Mahathir Mohamad shut down the Malaysian economy's links to the global markets in the midst of the Asian financial crisis, he was villified. But he understood very well the concept of the circuit breaker that prevents the transfer of risk.

Let's return to the workers we started with, those marginalised by the death of friction and removed from the value chain. In a landmark book, *The Brave New World of Work*, the German social thinker Ulrich Beck points to what he calls the individualisation of work. Until recently, developed societies have worked on models of large corporations with structured systems and expectations, where the burden of risk for an individual's future is carried largely by the employer or the government, or a mixture of the two. For the employer, the task is to attract good workers and retain them, progressing them where appropriate through their careers, and giving them pensions when they retire. Human capital pundits still think along these lines. For the government, the task is to invest in systems and structures for safety nets, because poverty is socially destabilising. In Europe, this results in strong social security systems, for example. Governments take a percentage of earned income, and invest it to provide medical, welfare and retirement benefits – all hedges against risk.

On the one hand, however, the pressures of the connected economy compel employers to remove humans and circuit breaking boundaries from their activity wherever they can.

On the other hand the flexibility of knowledge work and low personal technology costs allow people to be much more flexible in determining their own paths.

For differing reasons, therefore, the net result is a fragmentation and individualisation of work, with organisations taking less and less responsibility for the long term welfare of their employees. Individuals migrate in and out of formal employment, holding multiple roles, and often migrating across countries, constructing their own livelihoods wherever the economy offers them the possibility to do so.

The connected economy worker and risk

And with unstructured "grey work" on the rise, the flow of controlled employee earnings into government social security systems is drying up, meaning that governments too face the prospect of not being able to provide for the risks the future may bring. Greying populations who require heavy medical infrastructure support are a worrying trend in a situation where there are no longer structured mechanisms for catching a portion of their earnings while they are young. This is again about a transfer of risk. As friction decreases and workers become more responsible for their own futures, they are also acquiring the risk that employers and governments formerly held, and the ability of governments and organisations to take ownership of that risk is diminished. This is the darker, bleaker, fad-free side of reducing friction in the connected economy. Whenever risk moves, value is released. That is the meaning of paying an insurance premium. It is the rationale for creating complex checking, accounting and auditing systems.

But it now seems that the reverse is also true, that when value is released, risk also moves. In the hype driven marketplace of the new economy, nobody tells us this. In this latest, more radical movement of risk, we face two key dangers, only now beginning to emerge into view.

- First, the distribution of risk into the general population of consumers and workers is a division of risk. This is dangerous because at ground level it is very hard to recognise the quantum of the individual risk – until it hits us, or someone close to us. We don't get a sense of its true magnitude.
- The second danger is that we are not, as consumers and workers, educated in ways of dealing with this risk or providing against it.

Consumer protection groups are in some countries well developed, but in many others they are not – the Toshiba example quoted above is a graphic example of the power of North American consumers, and the consequent shifting of the burden of risk onto less powerful consumers in China. Worker collectives such as unions are another possible means of educating the new holders of risk, but while in Singapore they fill a critical educational role in the handling of new economy risks, in many other countries they either face losing battles on old economy battlegrounds (Korea, Germany), or they have been long defeated on those same fields (UK).

There is a chance that readers will agree with this whole analysis, and take a protectionist, Luddite stance against the brave new world of the connected economy. But there is no going back.

> Throughout the long summer of 1348, while millions died in Europe of bubonic plague, the ships and the caravans kept on moving, as they did year on year after that.

Commerce continues, and technology almost never retreats.

Who benefits?

The task that awaits us, therefore, is the task associated with all new risk. We need to learn about how this new risk works, on a scale comparable to the education in public hygiene that launched the twentieth century as the healthiest in human history. Specifically, we need to develop a new sense of the circuit breakers that reduce risk while retaining the new value flows of the connected economy. And we need to ensure that value released from the "death" of friction is deployed responsibly to include those who are most at risk of suffering from it.

> Frank Prestipino is Oracle Corporation's Director of e-Business for Asia Pacific. He tells the story of his first part-time computer operator job for a major beverage franchise in Australia, when he was just out of college. He noticed that the packing and distribution of the bottles on their way to customers was very inefficient. Each customer order was compiled on pallets at the warehouse and stacked in reverse order of delivery. But because the pallets were all of uneven size, they had to be stacked at an angle facing towards the centre of gravity of the truck. Hence less than optimal use of truck capacity. And because the customers could change their orders at the point of delivery, there was a lot of packing, unpacking and returns, involving both labour and paperwork. Frank designed a computer programme to overcome the evident frictions in this process. He had the different beverages stacked in cubes and slabs of 100 and 10 units. They were organised vertically in crates with different beverages in different positions, so that the truck operated like a mini-warehouse. The computer programme took the orders, and calculated the amounts of each product and how they should be stacked. The customer's order could now be compiled at the point of delivery, the truck carried many more items, and the system enabled a truck to increase its productivity from 15 journeys per day to 25. Frank was very pleased, and so were his managers.
>
> The unions, however, shut the gates and threatened to close down the plant as soon as they heard of it. The warehousemen were unhappy, because there was less stacking work for them. The drivers were unhappy because they had a bonus system that gave them $4 for every journey made, and they had more labour in stacking the customer orders on delivery. Loss of friction for the company meant lower value for the workers. The dispute was only resolved when new incentives were introduced to minimise the differences between the new and old ways of working.

Frank's system is now used worldwide. The old warehouse workers and drivers are long gone, and Frank has moved on in the world. Frank Prestipino's programme has released untold value for the corporation that employed him, simply by eliminating friction from the system. But the nagging question behind this story still remains: who really benefited, and why?

19
Buying Time: Coffee and the New Economy

Until very recently, we used coffee to speed us up. Instant coffee, invented in 1906, expressed the acceleration and transience of the technological age, and it was a fitting cousin to the machine-made cigarette – short sticks of temporary nicotine, coming to social prominence at around the same time. Originally a "poor man's cigar" most commonly used by soldiers (who, in modern warfare, didn't have the luxury of time required by pipes or cigars) the lightness and transience of the paper-wrapped cigarette came to symbolise speed, sophistication and modernity, just as instant coffee was to symbolise the enthusiastic, technologised acceleration of the American household: wife, gleaming kitchen, 2.4 kids and a husband driving daily to the corporate office to make the world go round – faster.

Two hundred and twenty years earlier, sometime around 1687, a man named Edward Lloyd opened a coffee-house in London's Tower Street. It was so successful that he moved several years later to Lombard Street, seat of bankers and merchants. Lloyd was not an innovator: his coffee-house was established almost 40 years after England's first coffee-house – but he knew how to choose his locations well. His establishment became the most popular venue for ship-owners, ship captains, insurance brokers and bankers – in short, anybody interested in international commerce.

Inky, muddy and bitter, coffee has a profound pharmacological effect – stimulating the imagination, quickening the reason, raising energy levels, and amplifying and enhancing social interactions.

> In 17th century Europe, coffee had been percolating as a magic brew out of the Middle East and into the salons of Europe for two hundred years via the coffee-houses of the Ottoman Empire. In 1616 the Dutchman Pieter Van der Broecke stole beans and bushes from the heavily guarded source of coffee, the Yemeni port of Mocha in the Red Sea. By the end of the century, the Dutch East India Company had established coffee plantations in Sri Lanka, India and Indonesia, destroying the Ottoman monopoly and providing a platform to introduce coffee drinking on a vast scale to the European markets.

The world's second most traded commodity

Coffee became the unmarked lubricant of the global economy, and coffee-houses the exchanges where information and ideas and innovations flowed between the key players. While transportation technology was the driver of the economy, whether it be maritime or rail technology, information about what was happening in the dispersed global markets was an essential commodity.

And coffee-houses provided the social environments for the flow of information, while coffee itself provided the pharmacological stimulation for this flow.

> Edward Lloyd and his successors sold coffee for 80 years, but it should not be a surprise that they recognised coffee's growing role as a lubricant in the economic information entrepôt. Edward Lloyd himself ran a news service called Lloyd's News, published intermittently from 1696. Over the next 100 years, the information service gradually became more profitable than the coffee-house itself – and in 1774 moved to the Royal Exchange to become what is now known as international insurance underwriting firm, Lloyd's of London – abandoning coffee altogether for the true currency of commerce – timely information.

Going back further in time, 200 years before Edward Lloyd, another generation of drinkers used coffee to play with time. The Sufi mystics used coffee as a ritual drink in the dhikr ceremony where celebrants recited prayers and incantations until they fell into a mystical trance. Sufi seers used coffee in these states of advanced meditation to break the barriers of time and see into the future.

For much of the 20th century coffee has been one of the great understated but critical components of the modern lifestyle. Taken for granted as an energiser, stimulant, and social device, it is in fact the world's second most traded commodity after oil. Over 400 billion cups are consumed every year. And it has played an almost invisible but central role in lubricating the friction of the global economy for 500 years: providing foresight to the Sufi mystics, information to merchants and investors, and speed to technological society.

And then in the last decade of the 20th century, with the blossoming of the New Economy, something strange happened to coffee. It multiplied in value, leapt several steps in the value chain and transformed itself from a supermarket freeze-dried dealer in acceleration into a luxury experience item combining taste, complexity, multiplication of price, and slowness.

Coffee and friction

Think now of the experience in the modern gourmet coffee-house: the sense of complex processes, the clanking and banging and hissing and sheer noise of it; the complexity of the names and types of combinations, the complications of choice. Think of the times when the gourmet coffee-houses and the heartlander coffee-stalls are most full – and think of what happens. Few deals are struck here, little economic information flows.

- Mornings, office workers steal a few moments of calm, slowing them down and preparing them for the rush of the day ahead.
- A second wave of housewives arrives hard on their heels to co-ordinate their leisurely start to the day.
- In the evenings, students hang out.
- On Saturday nights, youths linger with delicious slowness around tables laden with iced coffee, and swap banter with buddies and girlfriends.
- On Sunday mornings, young families take breakfast.

In the kitchens of coffee-drinking households, the complex range of devices for preparing coffee seems deliberately designed to slow you down, imparting some of the Sufi ritualism into the process of getting stimulation from caffeine. Coffee is no longer about acceleration, it is about friction, it is about slowing us down.

Now friction is not supposed to be popular in the new economy. On the contrary, the new economy is supposed to be about the abolition of friction. And here we are, not merely cultivating friction, but actually paying a lot more money for it. The affluent take that for granted, but it is still possible to encounter instances of surprise at the swiftness of this transition.

The coffee industry

> Sometime in 2000, I was standing in a queue behind an overseas worker, an Indian lad, who had clearly been sent by his mates to buy coffee for the gang. They had given him 60 cents apiece, and he had entered the first coffee shop he saw, Starbucks. His embarrassment at the actual price, $3.10 apiece, was contagious. In the end, the Starbucks staff member paid for them out of his own pocket.

In just five years in the United States, between 1993 and 1998, the value of the American coffee industry increased by a third – from US$13.5 billion to US$18 billion. Most of that value creation was the shift from speed, cheapness and simplicity (instant coffee) to slowness, complexity and expense (fresh coffee concoctions).

> By 1998, according to a Euromonitor International Report, global retail sales of fresh coffee were double those of instant coffee, making up two thirds of the retail coffee market. In 1978, almost a third of Americans drank instant coffee, according to the National Coffee Association. Now that figure is less than 10%, despite aggressive innovations in flavour and branding from incumbents such as Nescafé since the early 1980s.

It is no accident either that Seattle, the Mecca of the new economy, is also the New Coffee capital of the world. Home to Microsoft, Amazon, and more than 2,500 software and technology companies, it is also the home of Starbucks and countless other gourmet coffee bars. Founded in 1987, Starbucks is now opening a new store somewhere in the world every working day. In Seattle there is a coffee-house on every street corner, and there are 60 different ways to order coffee, not counting the combinations you can make between the 60 ways.

Coffee and coping in the new economy

The name "Java" used to signify an Indonesian island: now the name also envelopes the fragrance of Javanese coffee and the power of Java programming. There is even a Singaporean software company named after coffee – Mocha3D, specialising in interactive 3D solutions for websites. And yes, their solutions do slow you down on your passage through their websites – they are meant to.

To discover the reasons behind the strange transformation of coffee and to fathom its crucial and deeply felt role in suddenly slowing us down after 500 years of speeding us up, we need to go back to the mythical origins of coffee itself.

In *The Arabian Nights*, the beautiful Scheherazade entranced her husband the king with an intricate web of stories, one nested inside the other, full of fantasy and wit. And although, fearing treachery, the king had ordered her dead at the end of their wedding night, he kept her alive from night to night, because she held him suspended in mid thread. And so she moved skilfully from story to story, until three years had passed, and presenting him with their three sons, finally ended her telling and begged for life.

> One of Scheherazade's stories told of a goatherd named Kaldi, who could never get enough sleep. His goats, grazing high in the mountains, appeared full of energy, day and night. Some holy dervishes from a nearby monastery became interested in the boy's restless goats. They observed that the goats were grazing on a tiny red cherry that grew on bushes in the area. Tasting it themselves, they felt themselves overcome by a sudden surge of energy. Thus entered the miracle brew, coffee, into human history.

Scheherazade's stories bought her time, time for her husband to learn to trust her, and spare her life.

And coffee buys us time. In my earlier chapter *Fast Technology, Slow People*, I suggested that when information travels faster than human brains are built to process that information, we engineer devices to slow the world down so that we can cope.

Harassed by the rush of the electronically enabled workplace, we no longer want to accelerate ourselves to catch up. Nor can we slow the world of work down, because the competitive arms race forbids it. What coffee does for us pharmacologically, is to create an illusion of amplification in our leisure time, and coffee-houses provide us the means of amplifying our social experiences with our friends, while also surrounding ourselves with the sounds, aromas and lengthy processes of a distinctly non-electronic environment.

The early 20th century author Gertrude Stein pre-figured this eerily when she wrote:

> "Coffee is real good when you drink it, it gives you time to think... It's a lot more

> than just a drink; it's something happening. Not as in hip, but like an event, a place to be, but not like a location, but like somewhere within yourself. It gives you time, but not actual hours or minutes, but a chance to be, like yourself, and have a 2nd cup."

Scheherazade used stories to slow time down, to ensure her survival. We use coffee now for similar purposes to ensure our social and temporal well-being in an ever accelerating and challenging world. More interestingly, however, the importance of coffee as a device to create friction rather than to reduce friction, suggests for the first time the possibility that friction of some kinds might be a good thing in the new economy. I shall be exploring some of these manifestations of "good" friction in future chapters in this book.

20
Spending Time: An Anatomy of Outrage

For almost a month in Singapore, from January into February 2001, *The Straits Times* Forum page bulged under the onslaught of short, sharp and bitterly angry letters about the levels of service at Post Office Savings Bank (POSB) branches – and in every case, there was a queue at the bottom of the problem. Why, in a nation that lines up enthusiastically for everything from Robinson's sale to National Day tickets to "nasi lemak" in Hougang town; would queuing become a national debate – with questions in Parliament and oblique cautions from the Consumer Association and ministers about the need for banks to take more care of the small customer?

The public ill-will that surrounds DBS Bank (which acquired POSB in 2000) even today is a complex mix of emotions. While it was sparked most visibly by the change in the queue ticketing system, there were other friction-imposing events in the bank's absorption of the POSB acquisition – branches closed, embarrassing network collapses, increased charges.

In previous chapters I have explored the impact of friction *removal* in the new economy. Now it's time to turn the tables, and look at the *imposition* of friction in the new economy, and how we as consumers respond. There are four main reasons why human beings intentionally impose friction on other human beings and make them wait:

- For the expression of *power* – the longer I can make you wait, the more evident is my power over you
- For the extraction of *value* – I need to restrict the access points to a product or service, so as to collect value from you
- For the purpose of *coordination and control* – I need to make some people wait so that I can coordinate their movements and activities in a strategically important direction
- For the purpose of *measurement and planning* – I need to find out more about what's going on, so I need to check, ask questions, observe and analyse

There are other causes for friction, of course, that are not intentional: accident, chance and incompetence name a few. In this chapter, I want to examine the DBS Bank case from the perspective of the first two of these reasons: friction as the exercise of power, and friction as the extraction of value.

The expression of power

Making people queue has always been a mark of potency. The language of delay permeates distinctions of status in languages from Chinese to English. In Chinese, the characters that express waiting are also associated with the words for court, law, and rank (shou hòu, dài, deng). In English, you attend at court (from the Latin "attendare", to wait), you wait at tables, you retain staff (from the Latin "retinere", to hold back).

The conspicuous consumption of other people's time has been a symbol of status throughout human history.

> In 1219, the great Taoist scholar Chang Chun, aged 71, and only eight years before his death, began a five year journey across China to the Hindu Kush mountains, at the behest of Genghis Khan, just to have a conversation with him.
>
> Marco Polo was retained at the court of Kublai Khan for 14 years.
>
> Christopher Columbus had to lobby the Spanish court for 17 years before he got an audience to promote his voyage to the New World.
>
> The 17th century English essayist Francis Bacon listed delay as the chief vice of authority, along with corruption, arrogance and favouritism.
>
> Louis XIV exercised control over the aristocracy of France by requiring them to hang around at court and spend their money on lavish clothing and lifestyles instead of raising armies to jostle for power, as their rebellious forbears had been wont to do. If they were not at court, they didn't get any perks, sinecures or favours.

The French revolutionaries took a dim view of the "idle rich" but for King Louis, this particular waiting game was a serious political matter.

Modern times are not excepted. The power of the totalitarian state was articulated in bread lines throughout Russia and Eastern Europe. The bureaucratic state replicates this in complex and bizarre processes that require your attendance for days at a time for driving licences, passports, housing permits. The length of an immigration queue is a physical articulation of that country's attitude to its visitors.

> When former President Joseph Estrada wanted to gain popularity among working class Filipinos, he instituted fast track lanes at Manila Airport for returning overseas workers.

In times of economic hardship, it is the dispossessed who suddenly become visible in the streets, queuing for jobs, queuing for cheap transport, queuing for welfare, queuing for rations, queuing for healthcare, queuing for housing. In the world's first modern democracy, the 19th century American writer on etiquette Fanny Fern, wrote "Always keep callers waiting, till they have had time to notice the outlay of money in your parlors."

In the developing world, status through race, rank and money means fast-track exemption from queues. In the United Kingdom, the only way to get away from National Health Service waiting lists is to buy your way into private healthcare.

If we have a high opinion of ourselves, we dislike the public consumption of our time as an expression of power over us.

> In 1967, Richard Nixon paid a private visit to India. As a former Vice-President, and presidential candidate for upcoming elections, he paid a courtesy call on Prime Minister Indira Gandhi. She chose to keep him waiting for several minutes – clearly something he would not forget.
>
> When, in 1971, Mrs Gandhi paid an official visit to President Nixon, she, too was made to wait in a reception room at the White House, for the same amount of time – and then a few minutes more. When Nixon finally appeared he was cool and stand-offish.
>
> Nixon's Presidency marked one of the poorest in relations between India and the United States, shifting the Cold War into a new gear that threatened to destabilise the whole of Asia: a US-China rapprochement versus a new Indian-Soviet cordiality.

The Richard Nixon syndrome is probably what lies beneath much of the emotion in the anti-DBS Bank feelings of past months. For in the new economy, we have been schooled in the primacy of the customer, the shift of the balance of power from vendor to buyer, the doctrine of

"Customer is King". Now we can take our rightful place as lofty arbiter of our supplier company's time.

And from John Major's Citizen's Charter initiative of the early 90s in the UK, to the Singapore government's PS21 vision and its most recent attempts to reduce healthcare queuing times, even government bureaucracies are seeking to liberate their citizen's time, and return it to them to do with as they wish.

> The Prime Minister's Office in Singapore employs government officers full time to seek out rules and processes that are not necessary – and break them. Banks in the United States give cash handouts to customers who have to wait more than the target queuing time. Government offices the world over are installing queue ticketing systems to streamline the flow of people and make waiting time more palatable.

And so when DBS Bank decided to remove the queue ticketing system in POSB branches, it appeared to the consumer, emotionally speaking, as a counter-strike from the old regime, and an arbitrary exercise of power in the interest of power alone. Queues in POSB branches had already been getting longer, with fewer branches and a lot of people consolidating their accounts before new charges took hold. But this was the last straw. With the ticketing system, you could live with a long queue. You took your number, you worked out how fast the queue was moving, and you went to get a cup of coffee, or did a bit of shopping, or simply read the newspaper. Your time was still your own to dispose of. Then came the apparently arbitrary decision to make people stand in line, looking at the back of somebody's head for 45 minutes.

It didn't seem to help when DBS Bank told its customers that at an average of six minutes (soon to fall to four) they were now queuing for less time. The people who queue for four minutes or fewer are not the people who suffer or complain. Peak time customers still experienced formidable waits. It didn't seem to help when customers were encouraged to use phone banking (not reliable, said the letters), internet banking (what if I don't have internet asked a reader), cheque deposit boxes (no record of deposit, says a concerned customer), or ATMs (long queues there too says one correspondent). Angry people clearly felt they were being cynically manipulated into the use of technology in order to save operating costs (then why are you charging us for internet banking, and why can't you make it work, asks a reader). And customers pushed back, emotionally.

Contrary to popular sentiment at the time, DBS Bank likely did not intend to alienate its customers on such a large scale. Though its sensitivity to public sentiment was sorely lacking—in one issue of a newsletter for credit card customers in February 2001, it cheerfully recommended the merits of business networking by chatting with fellow attendees in the queue.

While DBS Bank's decision to impose friction might well have had subliminal assumptions about its ability to exercise power over its small customers, its primary goal in closing branches and changing the queuing system was probably motivated by the second main reason for slowing people down – the need to extract value for giving access to products and services. The second part of this chapter will examine this need.

Up to this point, I have looked at the public outcry over the creation of new friction for POSB customers, and interpreted it as a reaction based on assumptions about power. Now I want to examine the debate from the perspective of friction as a means for extracting value.

The extraction of value

You sometimes need to impose friction in order to extract value from giving access to products or services. Ticketing offices that give people access to transportation, fairgrounds, theatres and cinemas, are all fertile queuing grounds. There is a delicate balance here between the value you can extract, and the costs of giving access. If a fairground wanted to abolish queues, it could multiply the number of booths. But it could have so many ticketing booths that the costs of maintaining them would outweigh the revenues received. And when you have too many service points, restricting their number will release value into your bottom line – so long as you can make people queue successfully at the ones that are left.

There is another, apparently counter-rational dimension to this. The value principle also means that where we see a queue, we are conditioned to assume that there is a thing of value to be queued for, predisposing us to join that queue. The British, almost professional, enthusiasm for queues results in the quip that if you stand still in the street for more than a minute in any British town, a line will form behind you.

Hello Kitty frenzy in Hong Kong and Singapore was pragmatically linked to the investment value of collectibles, but visions of overnight queues also stimulate primal urges in us to acquire the object of such intense waiting. Department store managers are delighted when queues form outside the doors, though they may frown at weary queues at the checkout. The customer's antipathy to queues is not unambiguous – when we choose to, we will queue for a thing of value, and queuing, strangely, enhances its value in our mind's eye.

DBS Bank's reforms were attempting to bring the operating costs per customer and per process more into line with competitive consumer banking practice. When you scale your operations up into a much bigger entity, your shareholders expect – nay, require you – to demonstrate the economies of that scale. And it does take less time per customer if you make the customers stand in line. It does take fewer seconds per customer to walk to the counter from the head of the queue, than it does to spot your number, yawn, stretch, and stroll from the back of the lobby. And you don't waste time arguing with the customer who went shopping, came back after you had skipped his number, and is now trying to argue his way back into being served. You can run the same number of customers with fewer branches and fewer people if you can serve them faster, or if you can propel them into automated channels.

DBS Bank genuinely thought that by removing the queue numbering system it was saving time and dealing with queues faster. But it was shifting the burden of time onto the customer. It was saving time and releasing value for itself, but it was doing so at the customer's expense in personal disposable time.

In the new POSB queue, the customer thinks, my time is no longer my own. I have to stare at the back of someone's head for 45 minutes because the bank doesn't want to spend its employees' time in waiting for me. POSB is consuming its customers' time, and we don't like it one little bit.

Why don't customers perceive the issue from the same perspective as DBS Bank? Why do we interpret it as an arbitrary exercise of power, demonstrating our powerlessness as consumers? Why does it result in correspondents recommending defection to more consumer-friendly banks? Why, despite DBS chairman Dhanabalan's statement that the bank receives more compliments than complaints, and the great efforts DBS Bank and POSB staff evidently make, does DBS Bank have such a poor public reputation for service quality? In absolute terms, there are plenty of comparable environments where we are perfectly happy to be kept waiting, sometimes for much longer.

The offence here was an increase in friction with no perceived value to the customer in return. If average queuing times in the aggregate dropped, this was irrelevant to the loss of the personal disposable time hijacked by the new system. The value, on the contrary, devolved to the principal agent in this case, DBS Bank. The bank's failure to acknowledge the lack of balance in this relationship simply reinforced the perception that said the small customer doesn't matter. Words to the contrary appeared to be empty rhetoric.

Add that to the tide of history, the new economy rhetoric proclaiming the advent of the powerful customer, and the increasing ability to create competitive advantage through serving that power. In the United States, supermarkets are now deploying portable point of sales devices at peak shopping times, because research shows that customers will avoid a supermarket if they see check-out lines, and go somewhere else.

> When British Airways stated that it saw no economic advantage from economy class customers and would minimise the services and space accorded to them while maximising services to business and first class passengers, the public backlash reinforced the airline's already poor public image, and ultimately saw the departure of the CEO who had made such a rash public claim.

Hence the powerful popular perception of DBS Bank as a poor performer in terms of customer service is now difficult to contest or eradicate.

We have entered an economy of time. In this new economy, where information and knowledge are in profligate abundance, the poverty we feel most keenly is the absence of time.

Time is the currency of our lives now, and there is never enough. Our work-lives overflow far beyond a 44-hour week, and encroach on evenings and weekends. We count our moments more and more carefully. And we are beginning to resist casual expenditure of our time by apparently arrogant companies. Queues consume my time, and profit from it; and I don't like it. I will actively seek other alternatives, and this is dangerous for those businesses that still consume their customers' time.

Companies that have not yet woken up to the fact that time is a precious commodity in our eyes in this pressing, hurried economy, are extremely vulnerable to competition on the basis of customer time. Worse, value extraction by consuming customer time is interpreted as an abuse of power and leaves legacy resentments that are hard to erase. Revenge is sweet, and is articulated in disproportionate public outcry, festering networks of criticism, and popular support for the new kid on the block.

Consumer products that are difficult to understand or install; services that you have to take a day off work to wait at home for; slowing people down in order to persuade them to buy something from you, or while they are buying something from you. All these create resentments that competitors can and will exploit. New economy companies are by no means exempt from this risk.

There are networks of ex-Pacific Internet (a Singapore internet service provider) subscribers who claim that normal internet access has deteriorated in quality from the time the company launched broadband services in 2000.

It takes a week for SingTel Magix (a rival internet service provider) to do the paperwork on a new internet account for existing telephone subscribers – in one of the most technologically literate telecommunication companies in Singapore.

President Nixon's revenge on Mrs Gandhi was a petty one, but it was a strong statement of who held the greater power (*see story this chapter*). Extracting value from customers by imposing friction may turn out to have short-term advantages and long-term risk where shifts in the balance of power take place. If the customer is indeed crowned king, old scores are going to be settled.

21
Innovating – Inside the Box

In June 1997, Singapore Prime Minister Goh Chok Tong made a speech before Parliament presenting the government's vision for the 21st century. His speech put innovation explicitly on the agenda for his government's new term, and you can see that drive behind practically every government programme since 1997.

It has taken some time to build an infrastructure of support for innovation in Singapore. Few people would argue that the Singapore of the last quarter century has built its formidable reputation on building systems (from schools to civil service to government linked companies to impressive regulatory strength) that deliberately inhibit large-scale innovation. It is an oft-made mistake to suppose that there has been no innovation. Behind its careful and implacable extension of control over the levers of economic growth, the Lee Kuan Yew government was strikingly flexible, imaginative and canny, unafraid to break other people's rules.

If the full history of Temasek Holdings, which handles government investments, were ever written, its subtitle would deserve to have the word innovation included.

> Government linked trading company Intraco obtained leverage over critical infrastructure commodities at a critical period in Singapore's growth, by working in areas then out of bounds to other Singapore businesses.
>
> Singapore Airlines and the Port of Singapore Authority have spectacular records of determined, focussed and pragmatic innovation, building not only world-class status for themselves, but also making a critical contribution to Singapore's status as a global transportation and business hub.

And at ground level, the SMEs of Singapore have consistently exhibited the same canny, opportunity-hungry spirit of the Lee Kuan Yew government, not least in their navigation through the nets of control installed by that government. That so few of them have become global players owes as much to limited horizons, limited access to capital and limited regional markets as to lack of innovation. The question might rather be how so many of them have become global players. So what is the problem?

Where is the innovator?

The issue has not actually been lack of innovation in itself, but the range of its availability. The issue has been that in order to achieve rapid, sustained growth, innovation had to be focussed and limited through very few, highly directed channels. You could only afford a few innovators, put them in positions of power, and compel everyone else to follow.

This was fine in the old economy. When a new economy with unknown territories, opportunities and rules emerges, however, you need a lot more ground level, globally informed ideas. Entire systems have to be taken apart and rebuilt, and when you are competing with idea-rich, highly exposed, risk-experienced Europe and North America, you have to do it quickly. Suddenly you need many more innovators.

Singapore's second problem is that it built its education system to produce a majority of followers, and a minority of innovators. That's fine for massive, concerted, centrally directed growth. But at a time of rapid change and the need for multiple and continuous innovation across an entire economy, the supply is suddenly, painfully short. When an economic downturn comes, you can re-calibrate the cost of doing business immediately by turning down the employer's contribution to workers' Central Provident Fund. It is not so easy to turn up the supply of innovation by readjusting the education system. That particular production line is 15 years in the making.

> There is a scene in the Monty Python 1979 film "The Life of Brian" (banned in a number of countries for blasphemy) where Brian, who has been mistaken for the Messiah, steps out on a balcony above a Jerusalem street one morning to find a vast crowd waiting for his wisdom. He is exacerbated by his sudden and unwanted lack of privacy, so he attempts to reason with them. "You don't need to follow me!" He cries out. "No! We don't need to follow you!" the crowd obligingly roars back. "You are all individuals!" continues the reluctant messiah. "Yes! We are all individuals!" roars the crowd with one voice. A lone figure pops up above the melée to register his dissent: "I'm not!" he cries briefly, before he is pounced on and violently suppressed by the ardent, unified, groupthink "individuals" surrounding him.

That scene, classic Monty Python in its absurdity and cruelty, sometimes returns to haunt me in observing the growth of innovation enthusiasm in Singapore. The Monty Python crew would not be kind: "Yes! We are all innovators!" cry the civil servants and the teachers and the corporate men as they churn out the latest "new" idea picked up in a management book or guru-laden conference. "I'm not," whispers the lone dissenter, and he is quickly whisked away out of the public eye.

"You can't suddenly ask middle aged managers brought up in the Singapore education system to become innovative if they are not creative already," says Sim Wong Hoo, Chairman of the Technopreneurship 21 Committee.

Producing innovators

While innovation can be nurtured, and creativity skills can be taught, it depends on slowly built habits of mind and strong enrichment from peers and environment. It is not easy to grow, and cannot be turned on at the flick of a switch. Too many Singaporeans interpret the call to innovate as a call to action rather than a call to tolerance and recognition – and the Prime Minister's 1997 call to innovate was precisely in the context of tolerance for diversity.

Too many managers believe they are being innovative simply by changing the way they do things in line with the latest fashion. Too many organisations believe they are being innovative when they produce more new ideas than their competition. Quantity may be evidence of ingenuity, but the creation of value from innovation requires absolute focus on the external market, not upon idea scoreboards sitting on department noticeboards.

"You don't want too many new ideas," says Sim Wong Hoo again, "too many ideas will slow you down. If you want to create value, you have to manage innovation carefully."

The hard truth is that Singapore has cultivated a generation of followers rather than a generation of innovators. I have written earlier of the immense challenges the Ministry of Education faces in turning around an education system geared towards conformity and packaged knowledge, and orienting it towards a landscape of open knowledge and experimentation, innovation and the handling of new risk. If the teaching cohort is becoming rapidly younger, it is no accident: old economy heads find it very stressful to teach navigation skills for a landscape they cannot see, and innovation is not a concept that has been taught once it has been defined. Value-laden innovation is evident only when it occurs, and then is recognised by its environment. There is a lot of risk involved. Conformists and followers don't feel comfortable in innovation environments.

What hope then, for innovation in Singapore, and what should her agenda be in the coming years? On July 1st 1981, Sim Wong Hoo banked his future career on being innovative when he founded Creative Technology. In the next chapter I shall be looking at the key issues in managing innovation based on an interview with Sim, but the starting point for Singapore businesses at the beginning of the 21st century must surely lie with the themes of self-awareness, tolerance and lack of defensiveness.

"Innovation is not that difficult," says Sim.

> "There are a lot of bright sparks around, and you only need a few bright sparks. We Singaporeans are too careful, too rule-based, and when we have no rules, we are paralysed. But there will be a few bright sparks, and we should use them, let them sparkle, let them flourish, keep an open mind, even if we ourselves are paralysed. The greatest mistake is to dampen creativity when it surfaces. Don't shoot those people down."

Self-awareness is the key to all this, says Sim.

> "Only when you admit that you are not innovative, then you can start thinking about how you can use innovation and accept new ideas."

22 Innovation: Out of the (Juke) Box

When we think of innovation, the temptation is to think about creativity happening in teams, resulting in new products in the marketplace. When you do courses on innovation, you mostly get a lot of work on idea generation, and on creating team environments that support creativity. Very few people look at the complex chain of events that ensures creativity survives beyond the first brainstorm. In fact, creativity is only a very small part of true innovation.

Most of what makes up innovation is not the newness or smartness of the idea, but the quality that makes it contagious. Contagious to colleagues and managers, who see why it's a good thing; contagious to suppliers, partners and distributors, who have to take risks and change the way they do things, to help you bring it to market. Contagious to customers, who need to understand and accept your new product pretty fast.

And radical innovations are generally not contagious – human beings are creatures of habit. We don't like new things if there is any burden of change or understanding involved. We like novelty, and transient attractive improvements, but we are slow to adopt radical innovations; and when we as producers have invested large amounts of money in building a new product or service, we can't afford to wait for the market to wake up. The key to successful innovation is in knowing how to make a new idea supremely infectious. We need it to work like a viral epidemic. This chapter will explore four key strategies that together can make an innovation contagious:

- The Trojan Horse strategy
- The Hygiene strategy
- The Immune Response strategy
- The Ecology Manipulation strategy

The Trojan Horse strategy

Creative Technology's Chairman Sim Wong Hoo hands over the latest version of the Nomad Jukebox. "This is not just another entertainment product," he says firmly, "the Jukebox is a vehicle into the new economy."

I am taken aback. To the untrained eye, the Jukebox looks very much like a portable CD player – except I know it can store and play 2,500 songs. So how would a music player with a very, very large storage capacity qualify as a "new economy vehicle"? And why would Creative Technology design it to look like a portable CD player, if it is such a revolutionary device?

Sim is quick to explain. Behind the apparent familiarity of the form there are radical implications in what the Jukebox can do. You can talk directly to PCs, and digitize your CD music into the Jukebox. You can download MP3 files off the internet. You can create and navigate entire libraries of music. You can create playlist performances for any mood. So far so good, it's pretty neat technology. But is it legal? Isn't this technology pandering to the Napsterizing of the music industry, bypassing legitimate content channels?

"We are a partner in the Secure Digital Music Initiative," says Sim. "We are talking to the content owners about a whole range of new economy models for how music gets bought. For example, we could sell the Jukebox with a few thousand pre-loaded most popular songs from the major labels. You get a chance to preview them, then you have to pay – maybe $1 per song. So long as the process is convenient, painless and has plenty of choice, this could be a very good legal alternative to stealing music off the web. No downloading time for a start." Now this is a more radical vision.

"This is going to revolutionise how music is bought, listened to, distributed and created," says Sim. The whole way that music gets produced would change radically if most music were purchased song by song. The album method means that an artiste will publish a portfolio of songs, and you pay $20 for 16 tracks, of which you listen regularly to maybe four. Would people buy those 12 marginal songs on the Jukebox method, and what will that mean for the portfolio building efforts (and careers) of artistes? Will it mean that singers and songwriters become narrower performers of "mega-hits", eschewing more experimental stuff at the edges of their ouevre? Would marginal artistes become more marginalised, simply by being less visible? "Ah," says Sim, "that's not all. The costs of producing quality music are now also coming down dramatically. At $300 our own product Sound Blaster Live! will do stuff that used to get done in $100,000 studios. Authors and musicians will find it much easier to get into the marketplace – the only question is whether they can get visibility to the customer."

Small wonder that Sim finds the content owners and the record labels cautious. "The nervous ones are taking the old economy view. They believe that if you can choose song by song, you will buy fewer. The new economy approach says that people have a regular budget for their music-buying. If I spend $100 per year on music, I'm still going to spend $100 per year with the new method. I'll just have more songs that I like." "The record labels are interested in what we are doing," he says, "but they are taking a very risk averse view, and they want the technology developers to carry the burden of the risk – for example, by crippling our products with features that prevent cheating – encryption, security, and so on. But of course that makes things less convenient for the consumer, and the risk is that they won't buy our product."

Now it is becoming clearer why the Jukebox looks like a portable CD player. Radical innovations that threaten to transform an entire industry had better reassure us with a sense of familiarity.

> When Johann Fust took the world's first batch of printed Bibles to Paris around 1456, he was nearly killed by a rioting crowd who believed that a man with so many identical copies of the Bible must be working in league with the Devil.

The first printed books, for all that, were faithful copies of their manuscript predecessors. The untrained eye would find it hard to distinguish a hand-written book from a printed book for at least a generation after the printing press was invented. It would take 40 years before the modern accoutrements of the printed book such as title pages, running headers, tables of contents, footnotes and graduated typefaces began to emerge as quite distinct from manuscript writing.

Yet even in this innocuous disguise, printed books were despised by the book-buying markets for over a decade. It was the content owners controlling the manuscript book industry who were behind much of the anti-print propaganda (it was they who stirred up the crowds against Fust in Paris). As in the modern music industry, there were powerful vested interests involved – from book-binders, rubricators, calligraphers, booksellers and illustrators. They all felt threatened by a radical new technology that provided massive quantities of cheaper books outside the channels and marketspaces that they controlled. They managed to delay the tide of change for 10 years or so. It was not until after Johann Fust's death in 1466 that the price of manuscript books began to fall, and the fortunes of the printed book began to rise.

There are more recent examples of camouflaging a radical innovation to gain easier acceptance. When plastic stoppers were introduced in wine bottles, they were red. They were slow in getting adopted, and now you'll find that they are more often cork coloured, with marbling effects that resemble cork. It's an easier transition to make, it seems.

The Trojan Horse Strategy clothes a radical innovation in familiar clothes, just as the HIV virus mimics the key that fits two receptors on the surface membrane of human T-lymphocytes. This is the means by which the virus gains admission to the cell and is able to commandeer the cell's resources to replicate itself and spread.

This mechanism in innovation involves having a keen awareness of what the "receptors" in colleagues and managers as well as the target market look like. For an innovation to succeed, first the idea must contaminate many minds in the chain of production on its way to market. This leads us to the second strategy.

The Hygiene strategy

Just having an innovation virus that sticks in people's minds and infects them is not enough. You also need an environment in which ideas can spread and infect others. The Hygiene Strategy is, more properly, a strategy for removing hygiene. For innovation to contaminate lots of people, you need a combination of infectiousness (the idea is the right shape to stick and gain resources for replication) together with an unhygienic environment that gives many opportunities for infection. You want a lack of hygiene.

The trouble is that most large organisations are structured to be hygienic – they have systems and processes and fixed lines and forms of communication that effectively filter out strange new ideas. There's a good reason for that. Ninety-five per cent of a company's business is not about novelty, but about making things real and consistent in an habitual, routine way. Only routine gives you leverage out of your people and resources, and only that leverage will give you your margins on investment and capital. "Too much innovation can slow things down," observes Sim Wong Hoo. "Innovators fight for their ideas, and they are passionate. They don't give up easily, and you have to spend a lot of time arguing with them. Sometimes this is not in the best interest of the whole. Sometimes you have to tell them to shut up."

Hence, company history is littered with examples of genuinely valuable innovation that met dead ends in corporate silos, locked out of the transmission vector by over-hygienic corporate structures.

> In the 1980s, researchers at DuPont came up with a self-decomposing polyester called Biomax. Originally intended for disposable nappies, it was then tried out successfully with DelMonte, who wanted a bag to cover freshly picked bananas, and that would decompose as the bananas ripened. But the Costa Rican manager who took on the project had an unrelated fight with his superiors, and the project lapsed. Biomax is a great invention that still has no practical application.

So what are the characteristics of unhygienic organisations? In basic terms, there needs to be a good mix of innovation-related personality types, and they need to be talking to each other a lot. Organisations with high levels of cross-functional collaboration innovate well. So do organisations which support serial innovators by giving them access to the early adopter personality types who can marshal support behind good ideas, and the implementers who can make things work. Innovation almost always involves conflict – innovators tend to be intolerant of rules and systems, yet ideas cannot be brought to birth unless practical systems are implemented to make them work. There is always a dialectic between creativity and system in the innovative organisation. The organisation also needs to have a rich culture of trust and respect and open communication if it is to rise above strong differences of opinion.

Organisations where innovators have easy access to senior managers as possible sponsors, also innovate well.

> At GE, researchers who built a new type of digital X-ray machine that used no film and allowed for remote viewing, found no ready acceptance until they somehow managed to convey the idea to Jack Welch himself, who then took it up and championed it.
>
> Richard Branson is famously accessible to the staff of his companies, and the Virgin Group is widely recognised as possessing an innovative, adventurous culture. The total wedding service Virgin Bride was initiated when a junior Virgin staff member waited outside Mr Branson's home all weekend to seek support for the idea. He hired her to set it up.

Conversely, organisations which operate idea silos and do not have promiscuous communication and access strategies, not only fail to innovate, but also fail to retain innovator talent.

> In their book *Radical Innovation* (2000), a group of researchers from Rensselaer Polytechnic Institute in the USA found that from the leaders of 12 innovation projects studied, four people resigned, two threatened resignation and two were dismissed.

We have already mentioned personality types in relation to innovation – these are the personality types who, when teamed effectively, can move an idea from conception to production and thence to marketplace. But widespread infection also means having personality types in your organisation who simply go around spreading the ideas and creating a groundswell of acceptance and support.

Malcolm Gladwell, in his brilliant book, *The Tipping Point* (2000) identifies three personality types that it pays to have on your side in the infectious, unhygienic organisation. Gladwell starts with epidemics to find his personality types, and he moves to ideas.

The first personality type whom he finds influential in the transmission of disease as well as ideas is the **Connector**. The Connector is an inveterate socialiser and networker. The Connector loves to make new friends and acquaintances – the colleague who will start a conversation with the taxi driver, find out their life history and then stay in touch with them, is a classic Connector personality. The Connector knows everybody and is liked by everybody. If a Connector likes an idea, you can be sure that they will be an influential ally in gaining widespread acceptance throughout their social networks. But if they have the flu, send them home.

The second personality type is the **Maven**. The Maven collects information, and dispenses it liberally to anyone who will listen. The Maven always has that piece of information you need, can recommend the best place to go for a good deal on a car this week, and is an encyclopaedic font of advice and practical tips. The Maven too is universally trusted, because their information dispensation services are completely altruistic. People go to them for advice, and they will always give good advice.

> It was the fashion mavens of New York who were credited with the sudden, completely inexplicable reversal of fortune for the Hush Puppies brand of shoes in early 1995. At that point, Hush Puppies were seen as dowdy, old fashioned and out of date. Their sales were so low that their parent company was thinking of phasing them out. Then, with no marketing or sales initiative to account for it, an extraordinary shift in sales began: within 12 months sales rocketed from 30,000 to 430,000 a year.

Gladwell's third personality type is the **Salesman**. He is the charismatic persuader who makes you feel warm and appreciated as soon as he starts talking. He is the colleague who has an uncanny ability to tune in to what you are thinking, to whom you talk so naturally, and who is most able to persuade you to change your mind about something. Don't be misled by the title: the Salesman always has the interests of the listener at heart – that again, is why they are consistently influential.

Organisations that want to be unhygienic for innovation to spread not only need to recruit and keep Gladwell's infectious agents, but they also need to ensure that the ideas that are produced are the right ones for these agents to pick up and transmit. The Connector, Maven and Salesman all come with a serious health warning: they can oppose a change in just as influential a way as they can spread it.

In the first part of this chapter, we looked at how an innovative idea needs to be contagious within the enterprise if it is to see light of day. Two further strategies are designed to bring the innovation into the marketplace where it can create new value for the firm.

The Immune Response strategy

So far, so good. You have clothed your innovation in familiar clothing, and it has gained some measure of acceptance. You have made your corporate environment unhygienic, so that people get exposed to your infectious innovation easily, and the result is that you have marshalled enough support in the different parts of your organisation to move it towards production. Now you are ready to ship, right? Wrong.

"When we developed the Jukebox," Sim Wong Hoo says, "we had quite a challenge to overcome." The device may look like a portable CD player, but storing and accessing 2,500 songs poses quite different issues. How, for example, are you going to provide easy navigation and access to your songs, in such a way that the user can build their own personal music libraries? It was easy enough to give the specifications, says Mr. Sim, "The system has to be easy to use."

Sim picks up the theme of contamination to explain why this is not such a simple thing to achieve.

"When a team develops a product, they become contaminated by their own vision of what they have created. It's very hard for them to see it as the customer will see it. In terms of user interface, the designers, the R&D guys, and the marketing department – they are all contaminated by their own unique view of it."

"So just before we shipped the Jukebox, we gave it to senior management here to play with – and we forbade them to read the manuals. We wanted to see if it could be used intuitively." Sim pauses and half smiles. "Of course, they couldn't work out how to use it properly. We had to delay shipment for more than a month while we worked on simplifying the steps and adding in help screens. We discovered that it's not enough to innovate effectively inside the company. We also have to know how to innovate outside the company. And that means we have to know how to contaminate the customer with our innovation in as easy a way as possible."

The Immune Response Strategy is a critical stage in taking an innovation to market. Moving an idea from its inventor's head into a real world product or service is a complex process. The act of infecting the people inside the company and getting their buy-in does not in itself guarantee acceptance in the marketplace. By this stage, it's not simply an idea, but a product. The innovation virus has evolved a long way in this process. New infection potential in the customer marketplace means the product has to be tested on guinea pigs for immunity or an allergic response, if the Trojan Horse Strategy that you started with is still to remain valid.

Companies that fail to complete this test effectively can suffer embarrassing failure.

> Monsanto thought it had the perfect innovation in its creation of a completely new market in genetically modified seeds. All the bases were covered: lower risk of crop failure, greater productivity, ability to cope with poor climates, lower costs, longer shelf life. They were completely blindsided by the public response to GM foods, a response so severe that to survive it Monsanto had to make a humiliating withdrawal and were forced into a merger with pharmaceutical group Pharmacia and Upjohn.
>
> Motorola invested billions of dollars in Iridium, a satellite telecommunications system that saw only temporary relief in the pre-Y2K run-up, and finally had to be abandoned and wound up. Iridium phones were bulky, difficult to use, and couldn't be used indoors, where most people want to make calls.

"We also discovered that you have to keep things really simple if you are going to innovate successfully," continues Sim Wong Hoo. "Why did Personal Digital Assistants fail in the marketplace? They were too innovative, they tried to do too much. The Palm Pilot, by contrast, is very simple; it focuses on two or three things that people can learn quickly how to do. It's a very successful product. When we created the Phone Blaster some years ago, we were very creative, but we invented a monster. We did all the right things, we had brainstorming sessions, we had focus groups, we did lots of studies. We put two things together, the phone and the fax utility, because we thought that would be innovative. But it was too slow, too complicated to gain widespread acceptance."

"You know you have a hot product when everyone in the company wants it, and they can pick up how to use it very easily," concludes Sim. "With the Jukebox, everyone in Creative wants one. Maybe they are going to wait until it gets cheaper, or until it's smaller or lighter, but they all want one."

The Ecology Manipulation strategy

With any radical innovation, no matter how much you clothe your new thing in familiar clothes (Trojan Horse), no matter how effective you are at contaminating people inside your company (Hygiene), and no matter how good you are at testing for customer immunity before you launch (Immune Response), there will often still remain a discontinuity between your innovation and the outside world.

For Sim Wong Hoo, this is an extension of the principle of "innovating outside the company". "Making an innovative product is not enough," he says. "You also have to do a lot of work in enabling the market."

This process starts way back in the production of a complex product, before you even get to the customer, and there is a classic story about Sony's co-founder Akio Morita that illustrates the point.

After the Second World War, Sony was primarily seen as a cheap manufacturing base for American products. Morita's ambition was to establish Sony as a producer of own-brand products to rival the Americans, and he studied American technology fiercely. In the mid 1950s, he decided that transistor technology was the way to go, and he set the company on producing the world's first pocket radio.

This was not an easy path to take: Sony depended on its own network of smaller suppliers, and these suppliers saw their ultimate economic masters as the American companies that commissioned work from Sony. To Sony's suppliers, re-tooling their factories to produce components for an untested product that had no guaranteed market in the USA was a big risk. Morita had to work hard to get his suppliers on board. Then, in 1957, when the first pocket radio finally came off the Sony production line, it was slightly too large to fit into the average shirt pocket. Morita was undeterred. He immediately commissioned a line of shirts with outsize pockets that would take the new radios.

The lesson of the Ecology Manipulation Strategy then, is the converse of the saying about Mahomet and the Mountain: if your innovation doesn't exactly fit the world, then you must try to change the world to fit your innovation.

This is an immensely difficult task to achieve.

Through the late 1980s and early 1990s, Dutch technology company Philips deployed its best research teams and spent billions of dollars in developing CD-i technology, a technology that merged video, hi-fi, games and interactive teaching into one player connected to a television set.

The technology failed, partly because it broke Sim Wong Hoo's Principle of Simplicity: it was trying to do too much, and was too complex a proposition for the consumer to easily grasp. But it also failed because there was a dearth of applications or content available on the market to use with the technology. Philips tried to sponsor or create content to mitigate this, but it tried too little, too late, and the technology never gained market acceptance. Similarly, when the first wave of WAP technology hit the market, early adopters who bought WAP phones were quickly disillusioned by the lack of functionality and applications available – and their response slowed down the market, placing higher and harder expectations on 3G telecommunications functionality.

"When Creative Technology moved into PC-based DVD," says Sim Wong Hoo, "I had to talk to so many players in the field to get them to work with us. The fact that we succeeded, meant that we were able to move the market two years ahead."

Moving innovation out of the company and into the marketplace involves a lot of persuasion and collaboration outside one's own company.

"In today's new economy, if you look at the complexity of the technology, no single company can take care of the whole chain necessary for the product to succeed. With the Jukebox we have to talk to PC manufacturers, software companies, music and content houses, internet companies and artistes. It's not an easy proposition to enable the market when there are so many different agendas at play."

Sim has a graphic example from Creative's history to illustrate the importance of influencing one's ecological environment in one's favour.

"When the sound compression technology in MP3 began to appear in the market in 1998, we had a big debate within Creative Technology about whether we should jump on the bandwagon. You see, we had already spent a lot of money developing a 10:1 audio compression technology five years before, but we had to write it off as an expensive lesson. Nobody could figure out what to do with the technology, because it was too new, and there was no market for it. So we initially started developing Nomad MP3 players very cautiously. It was only when the Napster debate picked up that we realised music transmitted over the internet had evolved into a real market for the technology. When we realised that, it stimulated a whole new strategy for us: the move into personal digital entertainment. And that means we have to cover all the bases: not just audio technology, which was always our core strength, but PCs, website community support, hardware, software and applications, the whole works."

You need strong nerves when you venture into market enablement. "We don't think the market is fully enabled yet," admits Sim. "We think a hard disk digital audioplayer like the Nomad Jukebox really ought to be selling for about US$299. The cost of hard disks is what drives the retail price up, and that won't come down until we get a lot more retail volume. But the market moves so fast that we decided we had to move fast as well to get the brand presence in a market we are sure will emerge. We don't make money now on the retail price... but we hope we are anticipating and educating the market in the right way."

Ecology manipulation carries very high costs, and very high risks. Markets are notoriously difficult to anticipate in a planned way.

> When Bell invented the telephone, he intended it as a device for transmitting orchestral concerts to distant audiences, and never anticipated its role in stimulating one-to-one communication at a distance.
>
> When Edison invented the phonograph, he intended it as a re-writeable recording device for storing voice messages, much like the later dictaphone. Instead, it took up the role of storing the orchestral concerts that Bell had wanted to transmit by telephone.

Somebody had to invent the radio and the dictaphone to fulfil Bell's and Edison's original intents. The point is, that the market found its own use for the invention in quite novel and unanticipated ways. All the more then, the core principles of successful innovation remain:

- Make it communicable
- Make it consistent with what people know
- Make it simple
- Make it useful

For an innovation to succeed the unpredictable vagaries of the marketplace, it must be attractive enough to the unready consumer so that it survives long enough to find its final prosperous niche.

Creative Technology's Jukebox is a fragile creature: it presupposes a new economy model that has not yet matured, it presupposes the unlikely collaboration of conservative vested interests in the music industry, it prefigures a personal entertainment industry that has not yet emerged. But when all is said and done, it's still a jolly good way for you to carry your music library around with you, download and listen to your favourite songs at will. Only time will tell whether this particular innovation succeeds in getting out of the box and into the pages of technology history.

23
In Praise of Laggards

When he was a child on a farm in rural America, sociologist Everett Rogers used to observe how neighbouring farmers took completely different approaches to the adoption of agricultural innovations. Some would always be trying out the latest gadget, while others would be steadily resisting newfangled things and sticking firm to methods and equipment that had been in their families for generations. More interesting than that, he noticed that the farmers who were most innovative, were not always the most successful with their implementation of new technology. Indeed, it was often later adopters who made the new stuff work better for them.

Since the early 1960s, Professor Rogers has been at the forefront of what's called "diffusion of innovations" research – his definitive book of that title first came out in 1962, and is now in its fourth, magisterial edition.

In his study of how innovations spread, Rogers distinguishes five key types of personality, each of whom takes a particular approach to new ideas, from the adventuresome innovator to the sceptical laggard. For the past three years or so, I have applied Rogers' taxonomy in workshops on the management of innovation and change – because attitudes towards change are at root very close to attitudes towards innovation. It has become increasingly apparent in these workshops that understanding one's own personal innovation profile is key to the successful adoption of a team-based approach to innovation and change.

Rogers' research suggests that in any given population (whether national or organisational) you are likely to find stable proportions of each type. Generally speaking, innovators will make up 2.5% of the population, early adopters 13.5%, early majority 34%, late majority 34%, and laggards 16%.

The stability of these proportions in a population tells us two things. First, that adoption of a new idea or change follows a bell curve, starting with a small minority, and with the transition to the early and late majority (two thirds of the population) being most critical for widespread success.

Second, that most people are predisposed towards a mix of different innovation profiles – and depending on our environment, we will be tipped to play the most deficient role. In other words, there is a particular ecology for the adoption of change, a stable system where people play particular roles as that ecology requires. To take an example, my profile combines early adopter with innovator (*see table on opposite page*). In a conservative organisation I tend to exaggerate my innovator tendencies, while in a generally more innovative environment I play the role of early adopter.

Innovation ecosystem

Now if there is indeed a social ecology of adoption, then that suggests we have evolved this because it's good for us. It suggests that each innovation profile plays an important role in the innovation ecosystem. It suggests, for example, that we need laggards. And in fact, when you look at what each innovation profile does, you can see that wherever some people push rapid innovation and change, you always have checks and balances to slow it down and scrutinise it for possible risk. What then does this innovation ecosystem accomplish?

1. It is an *innovation filtering* system
The early adopter makes adoption judgements based on her perceived ability to lead successful change and enhance her reputation. The early majority think through the idea for a long time and make sure they can visualise how it will work in the local setting. The late majority seek the opinions of many people in their local networks and like to see it in successful use before adoption. The laggard will adopt only when it is proven to work and has been in widespread use for a considerable time.

2. It is a *practical implementation* mechanism
Pure innovators are generally not good at working out how an idea will work in practice, or whether it will work at all. The early adopter transmits the idea in an acceptable form to the early majority who start the first tangible work in figuring out how it can work in practice, where it fits in the scheme of things, and what processes will bring it to fruition. The late majority define the rules and systems that will make it standard practice, and the laggards by their rationalised resistance stipulate what constitutes quality and best practice.

Role	Upside	Downside
Innovator	Ranges widely for new ideas; has wide personal networks, socialises with other innovators; has lots of ideas.	Socialises poorly with local networks; no follow through on ideas; doesn't communicate effectively with non-innovators about new ideas; has too many ideas that don't work; needs early adopter to successfully communicate new idea to others.
Early Adopter	Likes to associate with innovators; has good reputation among peers and local networks; communicates new ideas very well to colleagues; is looked to for leadership in change and innovation; likes new ideas but judges their practicality, likelihood of success, and whether the ideas enhance own reputation.	Tends not to innovate well without innovator associate; can put reputation before adoption; needs supply of early majority profiles to get support for implementation.
Early Majority	Like to adopt new ideas early; highly aware of the needs of local networks, and think new things out very carefully before adopting them. Excellent for figuring out how the innovation will work in practice.	Can be slow in thinking through the adoption of a new idea, and will seek a lot of advice. Tend not to adopt very readily without leadership of an early adopter.
Late Majority	Do not adopt until a majority of people in the ecosystem have adopted (innovator + early adopter + early majority = 50%) – late majority thus represent the adoption tipping point. Relate well to established rules, systems, and processes, thus good at defining conditions by which an innovation becomes standard practice.	Tend to be sceptical and critical of innovations, can be seen as active resisters. They network well locally, so they can be influential opponents of change. If it isn't systematised and doesn't have clear rules and guidelines, the tendency will be not to adopt.
Laggard	Point of reference always the past; traditional ways of doing things. As minority, must articulate clear reasons for resisting change, hence provide very useful input on issues of process, quality assurance, and risk exposure arising from innovation. Though framed negatively, input valuable for making innovation durable and productive.	Strong tendency to resist change actively. However, they tend not to be isolationists, and don't relate very well to local networks, so they have less negative opinion leadership than late majority.

3. It is a *risk management* mechanism
Innovators have many failures. As in venture capital, there is a high proportion of failures to success, hence the returns must be large on the successful ideas. Each step in the adoption chain involves judgements, deliberation, cautious opinion-seeking and building of relationships to what is already known. The net result is to reduce the risk exposure brought by innovators to the larger population. In this light, where the innovator is the most active communicator of risk, the laggard is its most active opponent.

Team approach

The implications for the effective management of change and innovation are enormous. By instinct in an innovation or change project, we try to load up the team with innovator and early adopter types. Yet what the ecosystem approach tells us is that we need all five components of the adoption spectrum to assure us that the innovation will become truly pervasive. For example, we try to shut out the laggards because they are critical, sceptical, and actively slow us down. Yet if properly used, and if we would but listen, the laggards will tell us how to design an innovation so that it is ISO 9001 compliant even before it hits the production line.

The isolation of the late majority type is even more dangerous, because they represent a substantial portion of the population, they are well networked and can influence opinion, and they represent the tipping point of adoption from minority to majority adoption.

- Innovators: increased innovation, increased risk
- Early Adopters: leadership, vision, direction
- Early Majority: structure, processes
- Late Majority: rules, systems, outcomes
- Laggards: quality, reduced risk

It is worthy of note that we all tend to have a mixture of innovation profiles in our personalities. Usually we have contiguous tendencies: so an innovator may have shades of early adopter, or an early majority type may also have late majority tendencies. Sometimes, and these are more valuable instances, individuals have profiles which are at different points on the spectrum.

> In one case, a manager with an insurance company had strong tendencies towards early adopter combined with those of a laggard. When he had identified the presence of both sets of characteristics, he was able to understand why he had particular strengths in leading change in a very conservative, change-resistant environment. "I know exactly how the laggards feel," he said, "so I realise that I can explain the changes more effectively to them and get their support."

In the context of Singapore, and its self-conscious lack of innovation capital, there are also interesting lessons to be learned. Chairman of Creative Technology Sim Wong Hoo, in an interview on his approach to innovation in early 2001, described his approach in setting up new project teams.

"We often say that Singaporeans are very *kiasu* and risk averse," he said. "But this has its advantages too. Singaporeans can design very robust products to very high quality as a result."

So when Sim starts a new project, he will balance the team between American members "who provide all the wild ideas" and Singaporean members "who will make the ideas bulletproof". "This has worked very well for us," says Sim.

> "If we had a team made up only of go-getting Americans, we'd probably ship the product too early. If we had a team just of Singaporeans, we'd probably never get it shipped, because it would never be perfect."

In Sim's case, the interplay between creativity and discipline creates a unique competitive advantage. More importantly, it also acknowledges that the archetypal Singaporean character can play a critical role in successful innovation – because innovation is not simply about surfacing new ideas, but more importantly, about making new ideas real and getting them adopted.

I'm not sure that the Singaporean character is pure laggard. There are areas such as shopping, food and public administration where Singaporeans innovate constantly. But the attention to systems, pragmatic implementation and risk aversion are all characteristics of the conservative end of the innovation spectrum. It's time to acknowledge that these characteristics are valuable and honourable ones – if, and only if, they can be made to work well with the more creative ends of the spectrum. Laggards should not be avoided or denigrated. They play an essential role in making sure things work well, and in protecting us from the folly of premature adoption. Let's raise our glasses to the laggards within us, and give them a more positive voice.

24
Friction is Dead: Long Live Friction

> In book nine of Homer's *Odyssey*, our hero Odysseus arrives at the Isle of the Lotus Eaters. The crew he sends ashore to collect water and provisions do not return, so he sets out after them. Landing on the island he finds an idyllic place where beautiful women serve him food and drink... suspicious, he alone of all his shipmates refuses the honeyed lotus that is offered. Those who eat the lotus become langourous and slow, suddenly content to stay on the island and eat nothing but the delicious flower. Odysseus, strong minded, ties his fellow voyagers with rope and drags them to the ship, forcing them to leave against their will.

When I arrived in Singapore a decade ago, I planned only to stay for a short while. Like Odysseus' shipmates, I have stayed longer than I ever anticipated – but every two years or so, I have had the opportunity to consider deeply whether I want to stay or leave, and I have chosen to stay. That acquisition of growing influence over one's own destiny, once a privilege only of the wealthy, is a growing characteristic of the common knowledge worker in the new economy, and it is becoming more and more widespread.

There is a deep irony in the new economy's claims to abolish friction. While capital, labour, information and transactions move faster than they ever did before (not always at the pace or to the places that we would wish), we are suddenly discovering that friction is not always such a bad thing.

In previous chapters I have pointed out the unforeseen ways in which removal of friction can multiply and redistribute risk. Now we are also discovering that sometimes we'd like things to move more slowly – and people in particular. When it comes to customers, valued partners and talented employees, you want them to stay in your domain for as long as possible. And yet at the same time we are making it easier and easier for customers, partners and employees to slip away to other destinations in this lubricated information space we are struggling to call home.

Strong and weak ties

The key to this ease of movement is the visibility that the networked economy gives us. Before the Net, the realm of possibilities open to us (as customers, business partners or as employees) was restricted to the strength of our personal networks. For most people, these networks extend to family and friends and strong existing business relationships – what the sociologists would call strong ties. These networks are tight, relatively limited, and evolve slowly.

With the Internet comes a rapid proliferation of what the sociologists call networks of "weak ties" – looser connections, with lower loyalty, but extremely leaky in terms of knowledge and information exchange. What weak ties give us are much broader, wider perceptions into the world of possibilities available to us out there in the world – the more points in the network we communicate with, the more we see (sociologists again!) the "network holes" – spaces where needs are not being met, spaces that we might well perceive as opportunities.

Into this heady mix throw the rapid wash of capital around the world, flowing chaotically from one continent to another, seeking better leverage and better returns as markets become globalised and financial standards converge. Where money flows, opportunity springs up. Where opportunity flourishes, the news spreads through widely dispersed weak tie networks, and when such networks buzz, people pack their bags and leave. The world is seeing some of the largest shifts in population it has ever known, whether it be the legitimate flow of knowledge workers following global talent opportunities, or the illegal trade in human traffic, leveraging on the desire of poor people to go to a place of greater prosperity.

> An ambitious Kazakh worker doesn't need family or relatives now to find out how the labour market or immigration restrictions in Australia work; he just needs a cyber-café – and there is at least one he can go to. "The Internet is the only way I really have of knowing for myself what the outside world is really like," said department store clerk Nurlan in late 2000, interviewed by a Reuters reporter in Kazakhstan's former capital city Almaty. "There are people and ideas out there that no one here has ever imagined. I may never have the chance to see such things for myself, but the Internet sort of makes up for it."

Some people are not so relaxed as Nurlan professes to be about keeping their experience of the wider world merely virtual. Some people risk their lives for it.

Along with the highly touted achievement transmitting more bytes of information faster than ever before, sometimes it seems as though humans are trying to catch up. Human traffic is tragic when it involves displacement, overloaded and under-equipped boats, suffocating containers, exploitation and death. But more pervasively, the ties that bind us to our existing life relationships are loosening, we slip away from jobs, family, friends and homelands more easily than ever before... and more frequently.

> In a 1999 survey among more than 2,500 students in 11 countries, conducted by PricewaterhouseCoopers, almost three quarters of the students stated an interest in working abroad. Fewer than half expected to be working for their first employer for more than three years. Four separate Hay Group workforce studies recently found that fewer than 17% of employees surveyed were certain that they would stay with their current employer.

The new economy gurus have noticed this slipperiness among the human capital of enterprises, and given it the dramatic title "the war for talent" – only a few have noticed the irony of its cause. The business schools, consulting firms and technology companies have been advocates for years of policies that provide new channels for employee impermanence on the one hand, and actively drove employees out with a 10-year war against talent in the last decade of business process reengineering.

Now the race is on to make the organisation "stickier" so that talented employees will stay for longer – without any assurance of permanence to back it up. We want to be sticky for as long as we need you, the subliminal message says. The first, cruder approaches looked at incentives such as salaries, employee perks and stock options. The smarter ones have recently been looking at the notion of "social capital", the stock of trust-relationships, collaboration habits and communication cultures that an organisation possesses. Organisational culture is coming under ever greater scrutiny as a result, though most seasoned managers' eyes will glaze if they try to look at culture too long. It can't be counted, you see.

The difficulty in looking at culture, of course, is that it cannot discriminate so neatly between the people you want to keep and the ones you don't. Wages, perks and options can, but they seem to be remarkably ineffective in doing their talent retention job – as Lynda Gratton pointed out recently in her book *Living Strategy* (2000), tangible employment packages are much easier for competitors to copy than intangible, culture-based ones. Durable cultures such as those of Hewlett-Packard, Glaxo Wellcome and Motorola seem much more effective at attracting and retaining the talent they need than other organisations that slavishly copy the mechanical structures and systems they see there.

The consultants go hither and thither for new tools and new concepts to help...for a while during 2001, P2P (Peer to Peer) and "community-building" portals were all the rage, but worries about revolution and enthusiastic worker democracy swiftly transformed the focus onto B2E (business to employee) relationships. For all the jargon (or because of it), it's still not clear what those terms mean for the ongoing life and prosperity of the enterprise.

Talented customers, talented employees

In the first part of this chapter, I looked at the sudden realisation that we have reduced rather too much friction for our own comfort in our working environments. It is becoming harder and harder to retain good workers in our organisations. When it comes to customers, the lubrication of all the transaction channels has had an even more deliciously ironic effect – yes, you are helping customers move faster, but often you've made things so easy they don't get to perceive the value of your role in the process. Removing friction sometimes has the unwanted effect of accelerating a customer through your space and out the other side – if you're not careful, they may not even notice that you were there. Customer loyalty is hard to retain in an environment of surplus knowledge, brand intoxication and ever-wider options.

Customer Relationship Management has been with us now for a couple of years, and there is a growing maturity in the statements of what CRM might be – identifying your high value lifetime customers, and assuring that their encounters with you are pleasant, consistent, and maybe even delightful. There is a worrying lack of honesty here, however: there isn't much guidance on what you should do with your "low-value" customers. As with the notion of the "low value" employee, the gurus' eyes slide smoothly past these inconvenient questions. Let us assume that if high value customers and employees are to be retained, then low value ones must be discarded – as humanely as possible, of course. The implications are unspoken because they must be. There are deeply patronising and offensive implications behind the discrimination of "high-value" customer and employee capital. If we spoke about those implications, many of the very customers and employees we wanted would leave us.

The challenges of the networked economy, of course, are how to combine technology-enabled lubrication with a personalised, "friendly" experience for the valuable customer. Technology is not good at making us feel human, quite a lot of ingenuity is required to make it so. Few have succeeded in this enterprise, Amazon.com being a notable exception – and they have found it difficult to translate this success into large profit margins.

The truly depressing thing about this is that we knew all this stuff already. We are, after all, human beings. Every now and then a research firm will do some research that astounds us with its truth – because we ought to have known it in our heart of hearts, and yet have been playing abstract management games with models, charts and checklists, pretending that we were not human beings at all.

> In 2001 Hay Group published a working paper entitled *The Retention Dilemma: Why Productive Workers Leave*. In it, they found that between 60 and 70 per cent of the workforce feels that managers show no concern about their career and future welfare. They found very strong relationships between employee dissatisfaction and the lack of a clearly communicated sense of direction in a company. Dissatisfaction frequently went hand in hand with the belief among employees that managers were not competent and did not care about them. The failure to remove incompetent or weak managers was a strong demotivator.
>
> Strikingly, the report quotes Maria Grasso, Director of Organization Development at Dynamics Research Corporation: "People don't leave jobs, they leave their managers," she says. "I believe people would rather work for the

best manager at the worst company rather than the worst manager at the best company."

Why are we surprised at that?

The biggest challenge facing enterprises in this ever-lubricated, ever-technologised and ever-automated environment is the challenge of how to slow good people down while they are in your space, so that they can add value to it. It doesn't much matter whether they are customers or employees. The challenges are very much the same: how do you become attractive enough to talented customers and talented employees so that they visit your space, and sticky enough so that they will stay for a season and add value to your work? (You can hope for no more than a season).

I believe the identification of talented customers and employees is a more acceptable one than the identification of "high-value" customers and employees. Where the notion of value assumes a utilitarian view of a possession that you wish to exploit, the notion of talent allows us to speak of creating an engagement where value exchange takes place because of the interaction of talent.

Talented customers are those who involve themselves in you, challenge you and ask you for things. They take an interest, make suggestions and complain. They speak for the silent majority, they tell you things that will be true of the market in six months time. We believe we know what talented employees are, but too many of us still regard talent as a functional benefit rather than a benefit arising from passion and commitment in addition to the expected range of skills, competencies and knowledge. In fact, talented employees do very similar things to talented customers. They are often difficult to live with, and while we may often tolerate the "talented customer" we often relegate the "talented employee" to the out-tray, using the power that managers hold to ensure that this talent will never be deployed to our discomfort – or benefit.

How do you identify talented customers and employees so that you can call them over to take a look at you? How do you make sure you have enough likeable, talented customers and employees in the first place so that these new visitors will want to stay awhile? And how do you ensure that all your visitors feel that their life-goals will be honoured and furthered by you, even when they have left your space? "Attract and retain" goes the Human Resource mantra, harking back to the days of the corporate prison. "Attract and migrate" goes the new economy mantra, recognising that customers and employees will move on, indeed sometimes helping them to move on, but making sure they have a great story to tell others on their wider journeys.

For a while the knowledge managers sounded almost human when they said we should treat our employees like shareholders. Then we behaved like managers again and took the idea literally, handing out stock options and feeling we had done our part. Yet while we perform all our impersonal management tricks, we know in our heart of hearts that all we mostly want – and all that others want – is to feel wanted, trusted and valuable, in a place that is hospitable, where we have good relationships of trust and communication, and where there is a sense of growth, direction and possibility.

This is no Utopian dream. It allows for the complexity, confusion and difficulty of work in the new economy workplace. It allows for levels of stress and conflict and in fact assumes them. The search for humanity in the workplace is the only strategy we now have to deal with the endemic difficulties of work in constantly accelerating change. Yes, friction is dead, at the level of the transaction. At the level of humanness, friction needs very much to come alive, if we are to survive and prosper in the new economy workplace. And the skills and attitudes and basic humanity of the manager, at every level, are critical to this. Leadership in the new economy involves the new responsibility of the tough-minded Odysseus, who dragged his workers unwillingly from their dreams, and set them on the road to their individual destinies.

25
Tale of Dog and Duck

There is a cosmopolitan suburb of Singapore called Holland Village. For a time in 2001, it appeared to be living up to its name. Every morning for several months, as the clock approached nine, a hen would stroll past DeliFrance and Starbucks, heading for the Holland Road. This was no kampung chicken scratching for insects in the dust. It was a city chicken, taking its morning walk – you would not have been surprised to see a furled Business Times under its wing.

The phenomenon deepened after a month or so when a duck joined in the morning parade, always a few paces behind the hen. While the hen was poised, urbane and walked with an air of purpose, the duck seemed almost gauche, pausing occasionally at the window of Starbucks to gawk at the patrons inside.

The effect was an odd one. It gave breakfast in Holland Village a quirky air of sophistication, and yet it also bestowed a sense of calm. When a chicken and a duck stroll past your door every morning going about their own private business, the pressures of life seem somehow to be lessened.

One Monday at the beginning of August, all that changed. Earlier that morning, two dogs, a Jack Russell terrier and a golden retriever had slipped through an unlocked gate at their home almost a mile away near Jalan Jelita, and began a search for adventure in the direction of town. To observers in Starbucks observing the morning ritual of the serious chicken and the curious duck, there was little clue of the drama that was to unfold – until the birds suddenly fluttered in panic back the way they had come, excitedly pursued by the two dogs intent on sport, blood and the kill.

Dogs of profit?

After the dogs went a man who had leapt up from his morning coffee in Starbucks. As the yelping dogs pinned down the chicken and grabbed for its throat, the man seized them both, pulling them off and hurling them several feet away. The dogs seemed hardly to notice him and dashed again for the dazed bird. Feathers floated in the air as the man hauled the dogs off a second time, and an elderly lady turned up to rescue her pet. Deprived of their first prize, the duck was the dogs' next target. Here the chase was longer, because the duck knew how to fly, but didn't have the sense to fly far. Panic swelled in the Standard Chartered bank lobby when an unwitting customer let in both the duck and the dogs. A quick thinking bank employee grabbed the duck by its wings and held it aloft while the dogs were dragged away.

Half an hour later, the drama was over. The birds were safe at home, nursing a deeply bruised confidence. The panting, doggishly friendly dogs had been collected by an apologetic owner. And life in Holland Village returned to normal.

Not quite normal. The chicken and the duck no longer take their morning stroll through Holland Village. The attack on them was the result of a momentary lapse, and is unlikely to recur. But their confidence is dashed, and they stay at home.

They are not alone. "Business is not so good" says Mary Phang, who works at a drinks stall in Holland Village. "The economy is bad, everybody is staying at home." "People make the recession worse" says a taxi driver, "When Singaporeans stop spending money, they make everybody suffer more". He has reason to complain. Average taxi-driver income had dropped by an estimated 20 to 30 percent over the previous few months.

The drama that Monday morning in Holland Village was a striking metaphor for the despondency that grips us in these difficult times. If we, the poor consumers, are the duck and the chicken, then the forces of global capitalism are the dogs of merciless profit. We might be lucky enough to have a strong, protective government who will keep the dogs temporarily at bay, but the emotional force of the attack leaves us dazed, suddenly shorn of our previous poise and confidence.

Adversity school

Fortunately we are not chickens or ducks. There are whole generations of Singaporeans who were schooled in adversity. Few have had an easy ride since the prosperity of the early 1990s. The retail sector has had only momentary relief from economic pressures since 1997. The majority of small businesses have seen only glimpses of prosperity since the mid-1990s. The direction of the Singapore economy has moved crushingly and inexorably away from manufacturing towards services – and the pace is quickening. The construction sector has been sustained only by massive government sponsored investment at least since 1996. And yet more than half of the top 50 companies in Singapore in terms of performance, growth and profitability come from the manufacturing and construction sector. Yes, adversity creates failure and suffering. But it also cultivates strength and a hardy, indefatigable entrepreneurial spirit.

> At the end of 1996, a fight at a pop concert between Dayaks and Madurese immigrants in Borneo flared into open warfare throughout the province. While the countryside was systematically 'cleansed' of Madurese by Dayak warriors, the army surrounded the city of Pontianak with booby traps, mines and barbed wire. Dayaks who approached the city in pursuit of refugees were killed and their bodies burned. Official estimates of the death toll were in the hundreds. Journalist Richard Lloyd Parry, who visited the region six months later found witnesses who estimated the deaths to be in the thousands.

The picture he painted in his report "What Young Men Do" has a few curious elements. First, the Dayak campaign appeared to be spontaneous and yet completely consistent across the province. No mosques burnt, no looting, no attacks on government buildings. All Madurese found – men, women, children – were killed, beheaded, and their hearts torn out and eaten. Nobody knew who organised them, leaders seemed to spring up spontaneously. Second, none of the Dayaks interviewed admitted direct involvement in the killings, though their Madurese neighbours were being killed all around them. People spoke of Dayaks in trance, lost in a collective responsibility for their brutal acts. Third, the priests Lloyd Parry spoke to described a marginalised, increasingly disempowered people who had been consistently pushed to the boundaries by government policy and enforced settlement of the Madurese on the only thing left to them, their lands. The analogy of a swarm of bees springs consistently to mind. Nobody issues orders, everybody knows what to do in face of a threat.

The 1997 massacres in Borneo have been described as a case study in the sociology of evil, and Lloyd Parry's essay is a gripping testimony to the sudden power of evil to turn neighbours and friends into murderers. But they are also a case study in the sociology of collective action in the face of threat.

Adversity's graduate

The emergence of Singapore as a potent city-state at the heart of the Malay archipelago in a decade of political and racial uncertainty is a much more positive story of collective, coordinated response in the face of adversity.

The spirits of war summoned by the Dayak elders in late 1996 entered the souls of the warriors, removed their consciences, and gave them an all too literal thirst for blood. Such spirits care little for building human well-being, and they are blind to the future. Their retreat into the jungles of Borneo is an uneasy one, temporary at best until the fundamental problems facing the Dayaks are resolved. If those spirits represent the human potential for destruction, the spirit of resolve that built Singapore has a more lasting, creative potency.

Singaporeans habitually acquiesce in the criticism imposed from outside Singapore of being an unoriginal, compliant people over-directed by a 'nanny state'. Only one article of mine has attracted international press attention. A piece that I wrote about the challenges facing the Singapore government in adjusting its education system to foster more innovation was picked up by national and international newspapers and turned into the now-ritual repetition of the Singaporean stereotype. The criticism is too easy to be true.

> Yes, Singapore faces deep challenges in migrating towards a less directed, more innovative society. But the merits of Singaporean discipline are evidenced in the very history of Singapore's progress and prosperity. Innovation flourishes today in Singapore not only in food, architecture and shopping, but more fundamentally in creating stability and efficiency out of adversity and uncertainty. This is no technocratic construct. This is the product of a powerful, collective, co-ordinated national will, deeper and more pervasive than any small group of individuals can achieve.

More interestingly, even the strongest of leadership wills is powerless to create change unless the population accedes to – and itself wills into being – the direction that leaders set. The *panglima*, the Dayak generals who appear from nowhere in times of crisis are created by their communities in time of need. The directions they set express the will of the people who created them.

This is the difference between individuals and communities. Individuals will retreat in the face of overwhelming threat. A lapse in confidence in an individual results in sudden, sometimes deadly passivity. Strong communities generate leaders, ideas and action. Singapore has responded before with stupendous results, and will no doubt continue to respond decisively as new threats arise – while not forgetting the very human privilege of complaint.

In the midst of this gloom, we are reminded of our strengths. The most successful businesses in Singapore are among the industries most threatened. Adversity brings opportunity as well as threat. In the next chapter I take a look at one possible shape for that opportunity.

26
The Old New Economy

We are in the midst of economic depression, capital is in short supply. New economy firms, recently flush with cash, are collapsing, merging or being acquired. The industry speaks of consolidation, a winnowing of the weak from the strong. Massive capital investments in infrastructure and communications have continued unabated for a decade, and resulted in revolutionary new commercial possibilities, but also resulted in massive over-exposure and poor liquidity. The rules of commerce and productivity have been redefined, but it seems that the markets have overstretched themselves. Business falters. Suddenly the brave new economy is not so popular any more, and the pundits question that it ever had any validity. If this seems familiar, look again.

> The year is 1873, we are in the United States and it is Black Friday, September 19th. The investment banking firm Jay Cooke & Company has become over-exposed to railroad securities and it collapses. The market panics and sparks a depression that will last most of the decade.

Yet throughout this period, new technology continues to redefine the commercial landscape and B2C remote commerce establishes a foothold that will last until the arrival of the Internet.

> In 1861 the Western Union Telegraph Company had completed its transcontinental line by a combined policy of aggressive takeovers of small companies and new line construction.
>
> In 1869, the Union Pacific Railroad had joined east and west coasts at Promontory Point, Utah.
>
> The year 1872 was especially auspicious, despite the shadow of the Credit Mobilier scandal forecasting the liquidity crisis of the following year.
>
> In 1872, Western Union inaugurated the first electronic money transfer service, Montgomery Ward opened the world's first general goods mail order service, and the United States government enacted the Mail Fraud Statute, to be followed closely by the Wire Fraud Statute.

Three factors converged to ensure that the US economy expanded by almost 20% during the 1870s, despite the depression, creating a durable model for remote commerce: a robust continental transport and communications system; inexpensive and reliable logistics and fulfilment services provided by the Postal Service; payment innovations and a legal framework for enforcement and redress.

Trust and security – the key issues

Trust and security were, like today, enormous issues. Fraud in remote commerce was a perennial problem until well into the 20th century. Mail order scams included blackmail: respondents to an advertisement for sexual panaceas or potions might well prompt a Cash On Delivery shipment of the product at a very high price. If the recipient refused payment, he might be threatened with the publication of his unusual interests among the local community. Privacy was also an issue. Mail order companies habitually sold each other "sucker lists" of individuals who had purchased from them. When the promoter of the "Man Medicine" nostrum was jailed for mail fraud in 1914, the court ordered the destruction of a mailing list containing half a million names.

From our vantage point, the 1870s looks like a success story for electronic and remote commerce.

> In 1877, only five years after inauguration, Western Union's electronic money transfers were moving US$2.5 million annually in almost 40,000 separate transactions.
>
> By 1878, Montgomery Ward's mail order business was making US$300,000 annually in sales. By 1887 this had ballooned to US$1 million.

From the point of view of the players, however, life was anything but certain. Montgomery Ward was abandoned by his original partners when their stock was destroyed in the Great Chicago Fire of 1871. The Chicago newspapers were highly sceptical of his scheme and warned their readers of the risks of such a "swindle".

The expansion of the Union Pacific Railroad and Western Union alike were achieved at the cost of ruthless consolidation, takeovers, speculation, high risk manoeuvres and not a little irregularity.

> When Western Union completed its transcontinental link in 1861, it was one of 50 telegraph companies in operation. Only five years later, it had already purchased 340 companies, and by the 1880s, it controlled 80% of the country's messages. The so-called "Erie Wars" that pitched the robber barons Vanderbilt

> and Gould against each other for control of the country's railroads saw skulduggery on a grand scale. Many small players went out of business.

Massive infrastructure investments that balloon into irrationality have familiar patterns. The depression of the 1870s in the United States mirrored on a grander scale the aftermath of the Railway Mania in the Britain of 1845-7.

Hundreds of railway schemes had been set up in the early 1840s, sparking a speculation frenzy. By October 1845, 1,200 railways were being projected at an estimated cost of £560 million. The amount of outstanding railway liabilities was over £600 million, well above the national income. The London *Times* estimated that perhaps £20 million a year could safely be spent on the railways without starving the rest of the nation of capital. It was clear that something had to give, and it did.

> In October 1847, after the failure of more than 50 leading mercantile firms and four banks, the Bank of England itself was in danger of collapse, and was saved only by the government's suspension of the Bank Act. By January 1850, the value of railway stocks had declined from their peak by an average of over 85%, and their total value was less than half the capital already expended on them.

History replayed the ratios with a vengeance in March 2000. And yet economically, the resulting density of railway provision in Britain brought enormous benefits. Faster, cheaper transportation for passengers, raw materials and goods contributed to unprecedented growth. From 1850 to 1870, GDP was growing consistently five times faster than the growth in population. What lessons can be learned from the patterns of the past?

First, that radical infrastructure growth almost always inspires speculation frenzy, over-extension of confidence and capital, and eventual collapse, particularly among pioneers.

Second, that the shadows of collapse often conceal the workings of the eventual winners, who create advantage out of converging technologies rather than creating technologies.

Advances in infrastructure rarely take a step backwards. While infrastructure investment is often more expensive than anticipated and frequently too costly to sustain, failure provides entrepreneurial opportunities to the wise, ambitious and determined. In every 19th century economy you care to look at, the decade that saw greatest expansion of the railways also saw a Jekyll-Hyde partnership between economic instability and a positive step change in economic productivity.

The expansion of remote commerce in the United States despite economic uncertainty and fears about fraud demonstrates the key principles behind long term success. The points of friction for successful remote commerce in the 1870s focussed around trust, fulfilment, consistency, long-term credibility – as they do today.

Establish a reputation for trust

Montgomery Ward coined the phrase "Satisfaction guaranteed, or your money back", and he lived up to it. Sears Roebuck maintained a policy of refund and return if the goods were not up to expectations.

> Western Union already had twenty years of history as a reliable enterprise when it launched the money transfer service in 1872, and it became adept at managing its public image. By the 1930s its 14,000 distinctively uniformed messengers were as vivid a feature on the urban American streetscape as the brown uniform of UPS is today.

Manage risk

The early mail order companies managed risk through careful credit control. Customers placed orders with partial payment through post offices or telegraphic transfer, and payment would be made cash on delivery or at the local post office or catalogue store.

> Western Union deployed a complex and robust system of codebooks throughout its telegraph network to prevent fraud.

Authenticate your customers

Sears and Roebuck relied heavily on their clerks' knowledge of their customers to manage their risk. Customers would send in orders with a hand-written signature, and they would sign again for receipt of the goods. Credit-worthiness was relatively easy to handle on this model of personal familiarity.

Exploit convergence in infrastructure developments

For Montgomery Ward, the Postal Service and the railroads provided reliable delivery strongly supported by federal laws. Distribution was cheap, supported by government policy, and payment options were flexible. Exploiting the infrastructure also created revenue traffic that reinforced the position of the newly established infrastructural businesses, creating synergies that elevated the value proposition of every party to the collaboration.

Address customer needs

Rural farming and township settlements had expanded faster than commercial networks. Agricultural progress meant farmers needed access to bigger and better equipment, and they could afford more commodities as their farming exports grew in value. Meanwhile local general stores charged large premiums for limited ranges of stock. There was a strong demand for reliable, cheaper access to broader ranges of goods, and the expense of catalogue distribution and goods fulfilment was more than offset by customer demand and resulting opportunities for margins on scale.

Work within your clients' comfort zone

By 1872, the US Postal Service had played the role of communications provider and commerce enabler for almost a century.

> The Continental Congress appointed Benjamin Franklin its Postmaster General in 1775, and the first three holders of that post reported directly to Congress. In 1838, all railroads were designated as postal routes, from 1864 railway stations functioned as post offices, and the Postal Service introduced postal money orders in the same year.

By 1869, you could also remit money overseas with foreign money orders. When Montgomery Ward started his business, he was using channels of payment and transportation that had become predictable, reliable and essential elements of the average citizen's lifestyle.

Old issues, new opportunities

The issues of the new economy are not as new as we might think, but they do have different nuances today. The opportunities and threats arising from global remote commerce carry many of the issues of the last quarter of the 19th century: trust, legal frameworks, fulfilment, payment, consistency with the way we have always done things.

So let's take a Montgomery Ward pair of spectacles, and address the same issues he did in 1872. What are the opportunities for remote commerce in this new environment?

In the early 1890s, a rural farmer could order a sewing machine and a bicycle by mail order, enclosing a deposit through a money order purchased at his local railway station post office. His goods would take some weeks to arrive, but arrive they surely would, and he would pay the balance at his post office, upon taking receipt. Railroad, post office and merchant credibility conspired to wrap this transaction in trust and predictability, and make it one of the most typical features of American rural life at the close of the 19th century.

Now, at the beginning of a new century of commerce, in a new era of consolidation, and in a new atmosphere of innovation amongst adversity, the focus is again on making the new infrastructure work for the ordinary customer. Consistency, trust and established reputation are paramount. "Satisfaction guaranteed, or your money back" is all the customer ever wanted to hear.

Index

E

F

G

H

I

J

K

L

M

N

T

U

V

W